Oceans
and Coasts
An Introduction to
Oceanography

Oceans and Coasts
An Introduction to Oceanography

John A. Black
Suffolk County Community College
Selden, New York

wcb

Wm. C. Brown Publishers
Dubuque, Iowa

Cover photo: © David Muench 1985. Chapter opening renderings adapted from the cover image by Mike Meyer.

Once again—to Gail

Oceans
and Coasts
An Introduction to
Oceanography

Contents

7
Water and the Waters of the Sea   110

8
Biotic and Abiotic Relationships   126

9
Nutrients and Nutrient Cycling   148

10
Winds and Wind-Drift Currents   170

11

The Thermohaline Circulation: The Subsurface Currents 190

12

Coastlines and Coastal Processes 204

13

Estuarine and Intertidal Ecology 232

14

The Ecology of the Open Ocean 260

Boxes

Since oceanography is an interdisciplinary science, it must make use of the appropriate portions of geology, chemistry, physics, and biology in order to explain the sea's processes. Thus, a study of the waters of the world—fascinating in itself—provides a unique opportunity to introduce many of the major concepts of the traditional sciences in an interesting and innovative way.

This text is intended primarily for the nonscience major. No previous background in the sciences is presumed, and unnecessary terminology and math are kept to a minimum. Indeed, the most difficult math in the text is the simple division encountered in the density equation. There are, however, extensive discussions of the basic scientific principles as they relate to oceanographic processes; these, combined with the integrated approach and the coastal perspective, make this text appropriate for beginning science majors and those considering science as a possible major. In my experience an approach of this type often serves to identify a particular field of interest for both groups of students.

The majority of students are most familiar and identify more readily with coastal areas. Consequently I have attempted to build this text primarily around the processes that are closest to the students' experience—those of the coastal ocean. I feel that this is a valid approach, since all marine systems, from the deepest ocean to the shallowest bay, operate by the same or very similar chemical, physical, geological, and biological processes. Thus, a thorough understanding of these principles can readily be related to seemingly diverse areas.

A coastal approach to oceanography is also valuable, since students, either as a class group or individually, can easily observe coastal phenomena (if necessary, even a moderately sized lake can serve as an example). If the deep ocean is emphasized, actual observations of this sort are rarely possible, particularly on the undergraduate level.

I have also attempted to employ an integrated approach in the text. The many principles, theories, and other concepts are introduced at appropriate points and then are continually referred to in later sections in order to explain the pertinent oceanographic processes then being discussed. I feel that

Preface

this integrated approach is necessary so that the student does not lose sight of the forest for the phytoplankton. For example, the concept of density is explained and extensively discussed in chapter 6 and then is continually reinforced in later chapters to explain specific water movements, currents, etc. Similarly, waves and tides first appear in chapter 4, and then in later chapters are reintroduced to illustrate their effects on animal and plant distributions, coastal processes, and so forth. The result of this approach is that students do not lose sight of the major concepts, unlike approaches in which concepts are presented as isolated bits of information, rarely to be seen again.

Another feature of this text is a number of boxed inserts, written at a variety of levels. Some of the boxed inserts are historical; others discuss oceanographic methods and sampling techniques; some are for enrichment; still others are of a speculative nature. Some of the inserts also give recent findings not yet included in other texts; for example, the discovery of plants at new depths and the rift beneath the Rio Grande. Also included in the text are key terms at the beginning of each chapter and review questions and activities, a bibliography, and suggested further readings at the conclusion of each chapter. The numerous illustrations and photographs have been specifically chosen to demonstrate and reinforce concepts discussed in the text. The photographs include examples from both the East Coast and the West Coast.

As an aid to instructors in preparing and presenting class materials, an Instructors Manual and set of 32 acetate transparencies are available to adopters of this text. The Instructor's Manual contains an introduction, test questions, and additional activities for each chapter. The transparencies reproduce 32 two-color illustrations from the text, which serve as a handy supplement to classroom lectures.

In addition to those cited elsewhere, I would like to thank a number of people who gave their help in the preparation of this text. I am indebted to Ray Welch, Chairman of the Biology Department at Suffolk County Community College, who read innumerable drafts of several chapters. Undoubtedly his comments and help improved these chapters and, more important, clarified my thinking on several points. Moreover, as a sympathetic and understanding administrator, he also made my life easier in terms of academic duties during the preparation of the manuscript. Thanks must also go to Betty Deroski of the Chemistry Department for her help during the preparation of chapters 6 and 7, as well as to Margaret Harrison of the Porter-Gaud School of Charleston, South Carolina, for comments on several aspects of the text. Special appreciation and thanks must also go to Gail Marquardt for extensive editorial help and comments throughout the project.

I would be remiss if I did not mention the efforts of three special people from William C. Brown: Barb Grantham, Faye Schilling, and Karen Doland. Barb Grantham saw the manuscript through its final stages and is responsible for the overall appearance and layout of the text. Faye Schilling became involved in the project when I was completely stymied in obtaining appropriate photos and she generously agreed to help out. In reality Faye did much more than help out—she located the majority of photos that appear in the text and has my everlasting appreciation. Without Karen Doland I would have been tempted to give up the project at several junctures. Karen was always available to help with countless editorial details, questions, etc. Her help and expertise made the final stages of the project not only bearable but actually enjoyable. My only regret in finishing this project is that no longer, when I least expect it, will I get a phone call from these fine people. I can honestly say that I will miss working with them.

Harry Rooney and Harry White, also of Suffolk County Community College, must also be thanked— Harry Rooney for suggesting that I undertake this project and Harry White for overall support during the preparation of the manuscript. Mention should also go to Theresa Mary Rooney, whose christening I missed during one of the hectic "final" rewrites.

Marge Broderick and Linda Ferri deserve special mention and appreciation for the fast, accurate typing of the many drafts that led to the completion of the text. In addition Mary Stanisci provided valuable help in the final stages of this project. National Secretary's Week was, undoubtedly, established for these people.

I am also grateful to the reviewers of this text: Virginia Sand, Kent State University–Tuscarawas; Stephen G. Lebsack, Linn Benton Community College; Donald W. Clay, Arizona Western College; Robert C. King, San Jose City College; William P. Roberts, James Madison University; Paul R. Pinet, Colgate University; Neil Crenshaw, Indian River Community College; Michael C. Smiles, Jr., State University of New York–Farmingdale; Jeffrey Kassner, Town of Brookhaven/Division of Environmental Protection; Robert J. Stone, Suffolk County Community College.

John A. Black

Oceans
and Coasts
An Introduction to
Oceanography

1
Introduction

Key Terms

continental drift
density
plate tectonics
salinity
subduction
temperature

The sea is the most obvious feature of the earth's surface; approximately 70 percent of the surface is covered by water. Beneath this water are the familiar sands of the beaches, bottoms of bays, and the inshore ocean. Farther offshore this water covers a spectacular submarine topography of underwater canyons, trenches, mountains, and plains.

Unlike the continents, which are physically separated from one another, the oceans are continuous and interconnected. They extend from the Arctic to the Antarctic, separated only by small bits of land—the continents. It is, then, more accurate to think in terms of the world ocean, rather than of specific areas such as the Atlantic and the Pacific oceans.

Since the world ocean is continuous, it has similar characteristics throughout. In the 1870s oceanographers collected seawater samples from all of the seas of the world at a variety of depths. When analyzed, the samples were found to have quite similar characteristics, thus indicating that the seas are well mixed and interconnected. Further, several expeditions made it obvious that not only are the seas interconnected but many of their processes are also interrelated. These findings made it apparent that an integrated method of study was needed—scientists from many disciplines had to become involved in the study of the sea.

Oceanography: An Interdisciplinary Science

All oceanographic processes are governed by geological, biological, chemical, and/or physical principles. Since oceanography makes use of the appropriate portions of these sciences to explain the phenomena that are observed in the sea, it is considered to be an interdisciplinary science.

While it is customary to subdivide oceanography into geological, biological, physical, and chemical oceanography, these are artificial subdivisions. All oceanographic processes are interrelated, and it is impossible to truly understand one of the branches without a knowledge of the others. For example,

Earth: the water planet

Severe erosion caused by wind-driven waves. Many homes along this coastline are actually in the water at high tide. When erosion of this magnitude occurs, coastal processes become important to the home owner as well as to the oceanographer.

waves and currents are formed and move according to well-known physical principles and so are of interest to the physical oceanographer. When these waves and currents affect the distribution of marine animals and plants, they become important to the biological oceanographer. Similarly the effects of the waves when they break upon a coastline are important to the marine geologist or geological oceanographer interested in coastal processes, as well as to the coastal engineer involved in the control of erosion.

The sediments of the sea floor are generally considered to be in the province of the marine geologist. When these sediments interact with and alter the chemistry of the overlying water, they are important to the chemical oceanographer. When, in turn, the water chemistry influences the distribution of marine organisms, it becomes important to the biological oceanographer. And when these organisms remove some dissolved materials from the water and add others, the organisms and their life-styles become pertinent to the chemical oceanographer. In fact, chemical oceanographers would have trouble understanding their findings if the creatures of the sea were ignored. Obviously, then, individuals who deal with any given branch or subdivision of oceanography must be knowledgeable in all of the other branches in order to be able to understand and interpret their data.

Some General Concepts: Density, Temperature, and Salinity

Density, defined as the relationship of the mass of an object to its volume, is an extremely important concept in oceanography. An object's position in the water, materials in and on the sea floor, and even the position of the water masses themselves are determined by their density. A denser object will always sink beneath a less dense object (fig. 1.1).

Density may be summarized by the simple equation: $D = m/v$, where m equals mass and v equals volume. Generally the density of an object can be increased by increasing its mass while keeping its volume constant or by decreasing its volume while keeping its mass constant. Conversely, density can be decreased by decreasing the mass while keeping

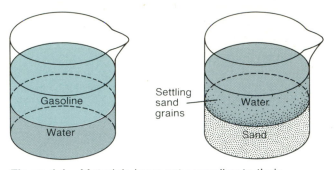

Figure 1.1 Materials layer out according to their density. Denser materials will always sink beneath those of a lower density.

the volume constant or by increasing its volume while keeping the mass constant. Consider, for example, a pillow that has a certain mass and occupies a specific volume. If one sat on the pillow, the feathers would be forced closer together and the volume occupied would certainly change; but the mass would remain the same. The density of the pillow would increase, since the particles that compose the pillow—the feathers—would be forced closer together and therefore would occupy a smaller space.

Changing the volume occupied by an object is, perhaps, the easiest way to change its density. This is customarily done by altering the object's **temperature.** For example, if the temperature of water is increased, the water will move about more rapidly and will occupy a greater volume; its density will decrease. If the water is cooled, the motion will diminish and the volume will also decrease; the water's density will increase.

It is also possible to alter the mass of the water by dissolving materials in it. The more material that is dissolved in a given quantity of water, the greater its mass and the higher its density.

Since the density of an object determines the position that it will occupy, a water mass having a high density will always move into and sink beneath one of a lower density (fig. 1.2). Basically this movement, in response to density differences, is the controlling factor of the currents that move deep beneath the sea's surface.

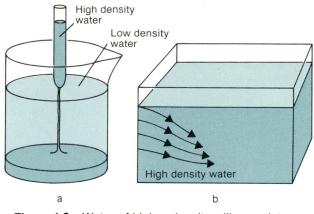

Figure 1.2   Water of higher density will move into (b) and sink beneath less dense water (a, b).

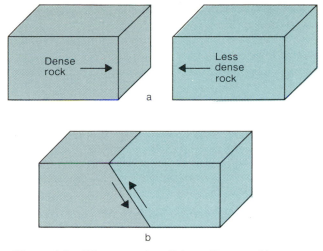

Figure 1.3   When a rock collides with one of lower density, the denser substance will move beneath the other.

The densities of the materials that compose the sea floor and the continents are also important. To illustrate, if a very large, very dense rock is pushed or pulled toward and into a less dense rock, the denser rock will generally be forced beneath the less dense material (fig. 1.3). As this rock descends, the friction may raise the temperature of these materials to such an extent that they will melt. The hot, molten material will expand and may come to the earth's surface as lava; a volcanic eruption may result.

Such an event did occur recently when huge portions of the ocean floor collided with the adjacent continent. The resultant molten material was forced to the surface and resulted in the eruption of Mount Saint Helens in the state of Washington. When the denser rock sank—that is, was **subducted** beneath the less dense rock—a deep ocean trench was formed. Again, it becomes clear that a knowledge of the composition and behavior of the deep-ocean floor is important not only to the geological oceanographer but also to geologists studying volcanoes hundreds of miles from the sea.

When oceanographers speak of the materials that are dissolved in seawater, they use the term **salinity.** Salinity is defined as the total amount of dissolved materials present in 1 kilogram of water. Since 1 kilogram equals 1,000 grams, the material is dissolved in 1,000 parts of water. For this reason salinity is stated in parts per thousand. Oceanographers use the symbol %oo to indicate parts per thousand when dealing with salinity.

The eruption of Mount St. Helens

Since the salinity directly affects the mass of water, it will also have a direct influence on its density. Thus, a seawater sample having a salinity of 20 %oo would be denser than one with a salinity of 15 %oo, assuming, of course, that the samples are at the same temperature.

Salinity is measured by using either chemical, physical, or electronic methods of analysis. A physical method, termed *hydrometry*, is one of the easier ways to determine salinity and relies on the density of the water.

An object such as a cork would be expected to float at different levels in water samples of different salinities, assuming that the temperature of the samples was the same. It would sink deeper into water of a lower salinity and float higher in water having a greater salinity. This is because the water with the higher salinity is denser and so will hold the cork up.

This same principle is used in hydrometry. A hydrometer is merely a weighted glass float with a long calibrated stem that looks much like a thermometer.

When the hydrometer is placed into a seawater sample, it will sink to a certain depth, dependent upon the salinity of the sample. One merely reads the number on the stem that corresponds to the water level and then uses a chart to convert the reading to the salinity of the sample.

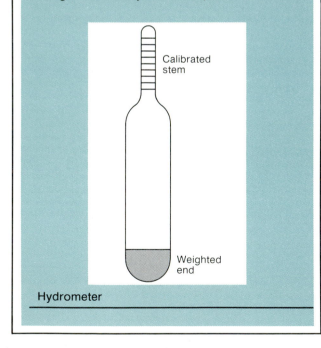

Calibrated stem

Weighted end

Hydrometer

In addition to influencing the movement and position of water masses, salinity also plays a major role in determining the distribution of marine organisms. Some organisms are able to tolerate large salinity variations and are widely distributed. Others are able to withstand only small changes in salinity and are less widely distributed or are confined to specific areas in the sea. These differences make the salinity of the world ocean very important to biological oceanographers studying the distribution of plants and animals.

Water temperatures are another important factor in the distribution of marine organisms. As is the case with salinity, some organisms are very tolerant of wide variations, while others are able to tolerate only a small temperature range. On a large scale the temperature of the sea is profoundly influenced by surface currents that sweep northward and southward from the equator and enable tropical and subtropical organisms to inhabit the more northerly and southerly, temperate seas. On a smaller scale shallow bays and other such bodies warm up and cool down much more rapidly than the waters of the deep ocean. These areas also receive large amounts of fresh water as stream- and river-flow and as runoff during storms. As a result the salinity is also very variable in these areas and the organisms that are successful in such bays are tolerant to wide salinity and temperature variations.

Wandering Continents

As noted previously it was the collision of the sea floor with a continental landmass that led to the eruption of Mount Saint Helens. Movements of this type are quite common. It was oceanographic investigations of the deep-sea floor that provided the conclusive evidence that convinced scientists that the sea floor and the continents are, indeed, moving about. **Continental drift,** or **plate tectonics,** actually serves to explain these movements and accounts for the present position of the continents and the topography of the deep-ocean basins. Moreover it allows one to reconstruct the shape of the earth's original landmasses and oceans (figs. 1.4 and 1.5).

Hagfish are unable to tolerate wide salinity variations. This fish has an interesting mode of feeding—it enters the mouth or anus of fish caught in traps or nets and eats them from the inside outward, leaving only the skins and bones. Hagfish are also called slime eels since they secrete large quantities of a slimy mucus when captured.

Parrot fish are restricted to the warm waters of the tropical seas. These fish have hundreds of teeth that are strong enough to enable them to scrape large chunks of coral from a reef when feeding.

Box 1.2 Latitude and Longitude

One of the major problems encountered by an oceanographer sampling the sea far from land is to accurately locate the sampling stations. This is usually done by using latitude and longitude. Both are merely grids of reference lines superimposed on the earth's surface in such a fashion as to intersect at right angles.

Lines of latitude begin at the equator, which is marked at 0° latitude. Additional lines are drawn about the earth at fixed increments to the north and south of the equator as far as the poles. The poles are marked 90° north and 90° south latitude, respectively.

Lines of longitude are drawn from pole to pole intersecting the lines of latitude at right angles. The 0° line of longitude begins at an arbitrarily chosen point to pass directly through the Royal Naval Observatory in Greenwich, England.

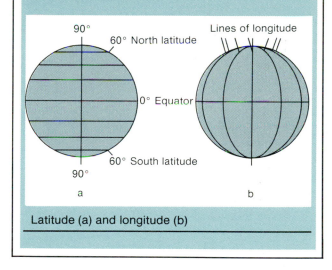

Latitude (a) and longitude (b)

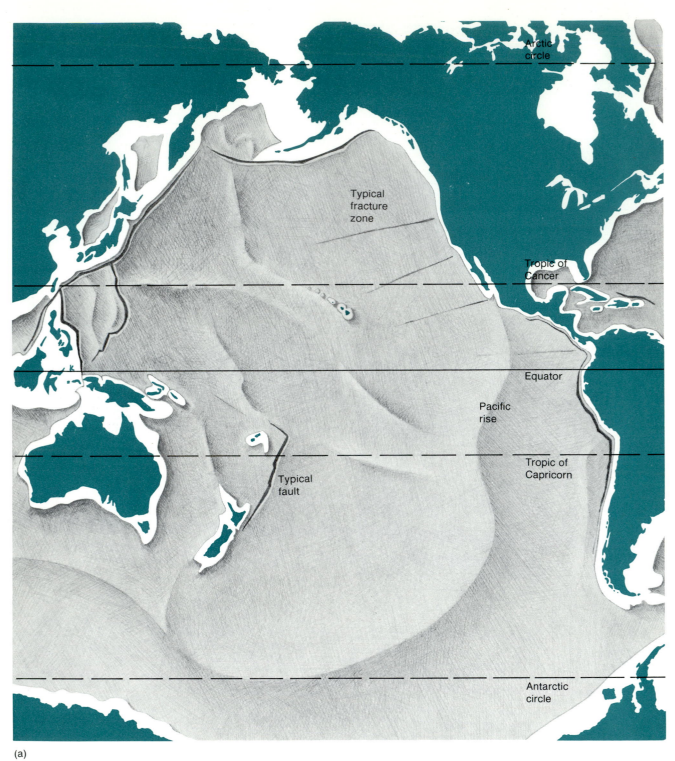

(a)

Figure 1.4 The sea floor of the Pacific (a) and the sea floor of the Atlantic (b)

Arctic
circle

Mid-Atlantic
ridge

Tropic of
Cancer

Equator

Pacific
rise

Tropic of
Capricorn

Antarctic
circle

(b)

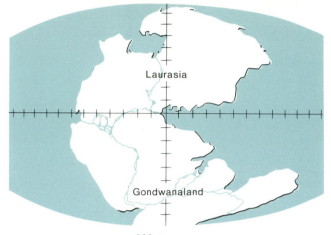

200 m.y. ago
Figure 1.5 The original position of the continents

	SYSTEM	SERIES	SOME ASPECTS OF THE LIFE RECORD	DURATION	YEARS BEFORE PRESENT
Cenozoic	Cenozoic	Pleistocene	Man	1,000,000	
		Pliocene		12 Mil	
		Miocene	Grasses become abundant	12 Mil	
		Oligocene	Horses first appear	11 Mil	
		Eocene		22 $\overline{M}$	
		Paleocene		5 $\overline{M}$	63,000,000
Mesozoic	Cretaceous		Extinction of dinosaurs	72 $\overline{M}$	
	Jurassic		Birds first appear	46 $\overline{M}$	
	Triassic		Dinosaurs first appear	49 $\overline{M}$	230,000,000
Paleozoic	Permian			50 $\overline{M}$	
	Pennsylvanian		Coal-forming swamps	30 $\overline{M}$	
	Mississippian			35 $\overline{M}$	
	Devonian			60 $\overline{M}$	
	Silurian		First vertebrates appear (fish)	20 $\overline{M}$	
	Ordovician			75 $\overline{M}$	
	Cambrian		First abundant fossil record (marine invertebrates)	100 $\overline{M}$	600,000,000
Precambrian rocks (abundant, but worldwide subdivisions not generally agreed upon)			Scanty fossil record — Primitive marine plants and invertebrates — One-celled organisms		

Vertical life-record bands: Flowering plants, Mammals, Conifer and cycad plants, Reptiles, Spore-bearing land plants, Amphibia, Fish, Marine invertebrates, Marine plants.

Figure 1.6 The geologic time scale

Plate tectonics also provides a model to account for the distribution of the animals and plants that inhabit the present-day continents. Since these movements have occurred over long periods of time in the earth's history, it is useful to refer to the geologic time scale (fig. 1.6) when studying plate tectonics.

Summary

Density, salinity, and temperature are very important concepts in the study of oceanography. The salinity and temperature of the water influence its density, and the differences in density are the major factor in understanding the formation of currents and the positions of water masses in the sea. In addition temperature and salinity play major roles in influencing the distribution of plants and animals.

The density of the materials that compose the sea floor and continents determine which materials will be subducted should collisions occur. Plate tectonics explains why these movements and collisions do occur and is central to the understanding of the earth's present structure.

Review

1. Why is it necessary to take an interdisciplinary approach when studying the sea?
2. Explain the importance of the temperature and salinity of the sea.
3. Define *density*. Why is density important?
4. What factors influence the density of a substance?
5. How is a knowledge of marine geology useful in explaining the eruption of Mount Saint Helens?

For Further Reading

Topic	Chapter
Currents	10, 11
Density	6, 7, 8, 10, 11
Plate tectonics	2
Salinity	7, 8, 11, 13
Temperature	6, 7, 10, 11, 13
Tides	4
Waves	4

2
The Continents, Continental Shelf, and Ocean Basins

Key Terms

active continental margin
asthenosphere
continental crust
continental plate
continental wandering
converging boundary
crustal rebound
diverging boundary
dust-cloud hypothesis
isostatic adjustment

lithosphere
magma
magnetic reversal
mixed plate
ocean vent
oceanic crust
oceanic plate
passive continental margin
plate boundary
transform fault
transform-plate boundary

The topography of the earth is remarkably varied. Mountain ranges, hills, plains, and canyons are present on the land, as well as deep beneath the sea. Although most of these features were formed millions of years ago, the mechanisms that formed them are still going on. As a result these features are constantly changing, and millions of years from now the continents and ocean basins will be very different from what they are today. For instance, when Mount Saint Helens exploded, the eruption significantly changed vast areas of the landscape in a very short time. In the process a large surrounding area was stripped of all of its soil and hundreds of kilometers of woodland were destroyed. This eruption was actually the result of the sinking of a portion of the ocean floor off the coasts of Oregon, Washington, and Vancouver Island, British Columbia.

The eruption of Mount Saint Helens graphically illustrates the fact that many terrestrial features and phenomena are due to processes that are occurring in or on the ocean floor. These processes are also responsible for the present position of the world's landmasses. The spatial relationships of the continents were quite different in the past and, since these processes are continuing, the continents will be spatially very different in the future.

The Formation of the Earth

One attempt to explain the formation of the earth and all of the planets in the solar system is known as the **dust-cloud hypothesis.** According to this hypothesis, the planets were formed from a diffuse cloud, or clouds, of dust and gas. Gravitational attractions between these particles caused them to come together in certain localized areas within the cloud. As these materials coalesced, they became more compact. This increased their mutual gravitational attractions and caused additional particles to be attracted to, and accumulate in, the forming masses. The process of attraction and the subsequent collision of particles within the embryonic planets served to increase their temperature. Thus, as the earth and the other planets formed, they were very hot—perhaps in a molten or semimolten state.

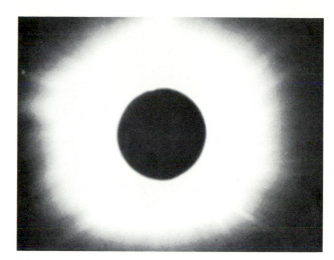

The gaseous nature of the sun viewed during a solar eclipse.

As the mass of the planets increased, those bodies became more strongly attracted to the sun and eventually formed well-defined orbits about it.

The sun, with a diameter of 1.39 million kilometers (865,000 miles), is so hot—with temperatures ranging from 6,000°C (11,000°F) at its surface to an estimated 20,000,000°C (36,000,000°F) at its center—that it is completely gaseous and self-luminous. Since it is self-luminous, it is considered to be a true star. All the planets in the Solar System (fig. 2.1) orbit about the sun in concentric, elliptical, counterclockwise orbits, as viewed from the North Pole.

The earth, with a diameter of 13,000 kilometers (8,000 miles), is 150 million kilometers (93 million miles) from the sun. It rotates on its axis while orbiting about the sun. The earth completes one rotation per day and one orbit about the sun per year (figs. 2.2 and 2.3). The orbit about the sun, combined with the inclination of the axis of rotation, is responsible for the warming and cooling of the earth, conditions that produce the seasons in the temperate and polar regions. When the Northern Hemisphere is inclined toward the sun, the sun's rays strike

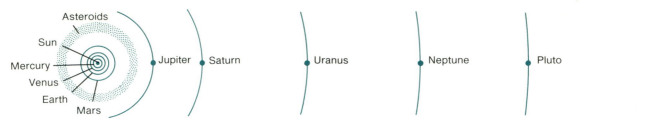

Figure 2.1 The arrangement of the Solar System. The orbits of the major planets are shown.

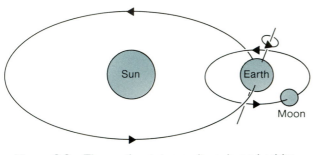

Figure 2.2 The earth rotates on its axis and orbits about the sun while the moon orbits about the earth.

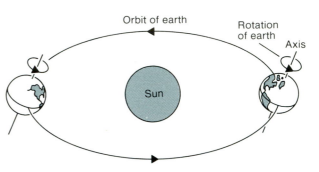

Figure 2.3 Perspective of the earth's orbit about the sun and its rotation on its axis. On the right side of this diagram, the Northern Hemisphere is tilted away from the sun and is experiencing winter. To the left it is tilted toward the sun and it is summer.

that portion of the earth most vertically, increasing solar heating and causing that hemisphere to experience summer. During the winter the Northern Hemisphere is on the opposite side of its orbit and tilts away from the sun. The sun's rays then strike at a greater angle, resulting in less solar heating and cooler winter temperatures.

The moon is much smaller, with a diameter of only 3,500 kilometers (2,000 miles). It orbits about the earth once every 29.5 days at an average distance of 390,000 kilometers (240,000 miles). As the moon orbits the earth, the earth rotates on its axis and also orbits about the sun. Thus, one side of the earth faces the stationary sun, but the other side need not necessarily face the orbiting moon. The positions of the sun, moon, and earth are important factors in the formation of the tides (chapter 4).

As noted previously, when the planets formed, their temperature increased because of the repeated collisions of coalescing particles. In their molten or semimolten state, the dense materials that composed the embryonic planets sank toward the central core, the least dense rose to the surface, and those materials of intermediate density moved between the surface materials and the denser ones.

As the earth formed and cooled, this process, termed segregation by density, resulted in a definite stratification, or layering. The denser materials, such as iron and nickel, sank toward the core, while the less dense material layered out above the core to form the mantle. Above the mantle the basalts and granites formed the surface crust (fig. 2.4). Since the basalt is the denser of the two, it is found beneath the granite of the continents. Basalt, however, does form the true ocean floor, although it is often covered by sediments.

The Structure of the Earth

On the basis of its stratification, the earth can be divided into the central core, the mantle, and the surface crust (fig. 2.5). The core actually consists of a solid inner and a molten outer portion. Both the inner and outer core consist of iron, nickel, and sulfur. The mantle, which is located above the core, is divided into the inner mantle and a softer, more fluid outer portion termed the **asthenosphere.** The asthenosphere is also known as the seismic low-velocity

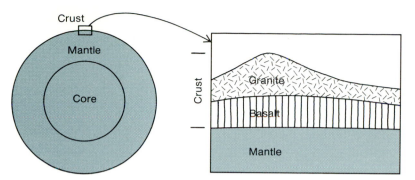

Figure 2.4 The relationship of the core, mantle, and crust. The crust consists of both granite and basalt. Since the granite is less dense, it layers out above the basalt and forms the continents. Basalt comes to the earth's surface on the deep sea floor.

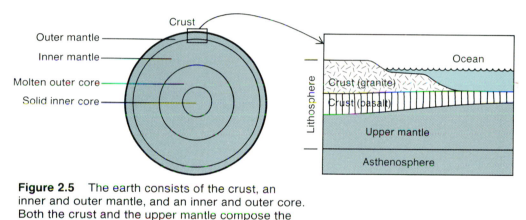

Figure 2.5 The earth consists of the crust, an inner and outer mantle, and an inner and outer core. Both the crust and the upper mantle compose the lithosphere.

zone, based on observation of the waves of energy that are generated by earthquakes. These waves, known as seismic waves, slow down as they travel into and through this portion of the mantle.

It is believed that the rocks in the asthenosphere are close to their melting point. This, then, may be the zone where molten rock, called **magma**, is generated and it may be the origin of the magma that comes to the earth's surface or to the surface of the ocean floor when volcanoes erupt. It is to be noted that the asthenosphere is located deeper below the earth's surface beneath the continents. There it is below both the granite that forms the continents and the underlying basalt. It is located closer to the surface when it is covered only by the basalt that composes the ocean floor (fig. 2.6).

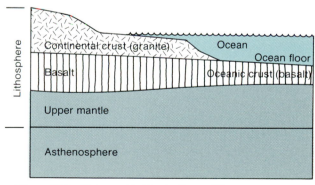
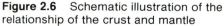

Figure 2.6 Schematic illustration of the relationship of the crust and mantle

When magma reaches the earth's surface it is called lava. This photo shows fountains of lava erupting from a Hawaiian volcano.

The **lithosphere** lies above the asthenosphere. It is between 60 and 100 kilometers (40 and 60 miles) thick and includes the **continental crust,** which is composed of granite and the basaltic **oceanic crust,** as well as the upper part of the mantle that lies above the seismic low-velocity zone.

The lithosphere can be thought of as "floating" upon the more fluid asthenosphere, which is soft and capable of flowing. If sediment is removed from a portion of the crust, the crust will become lighter and rise higher above the mantle, whereas if material is added to the crust, it will become heavier and sink deeper into the mantle. If material is neither added to nor removed from the crust, the crust will reach equilibrium and then is said to be in isostacy.

When the equilibrium is disturbed, by adding or removing material, the crust will move in order to restore this equilibrium. This movement is called **isostatic adjustment.**

During the glacial periods, the weight of the huge quantities of ice depressed the underlying crust. After the glaciers melted, the crust began to undergo isostatic adjustment as it moved upward in response to the decrease in weight. This process, which is known as **crustal rebound,** is still occurring in some areas (fig. 2.7) and is important in interpreting submerging and emerging shorelines (chapter 12).

The raised sea stack and marine beach
(background) provide evidence of isostatic
adjustment along Newfoundland's west coast.

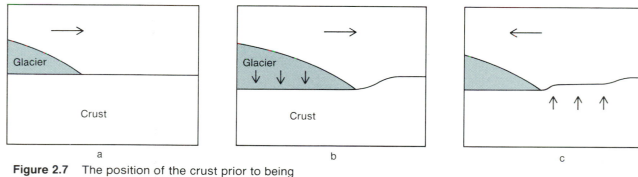

Figure 2.7 The position of the crust prior to being
covered by a glacier (a). The weight of the glacier
depresses the crust (b). After the glacier recedes
the crust slowly rebounds (c).

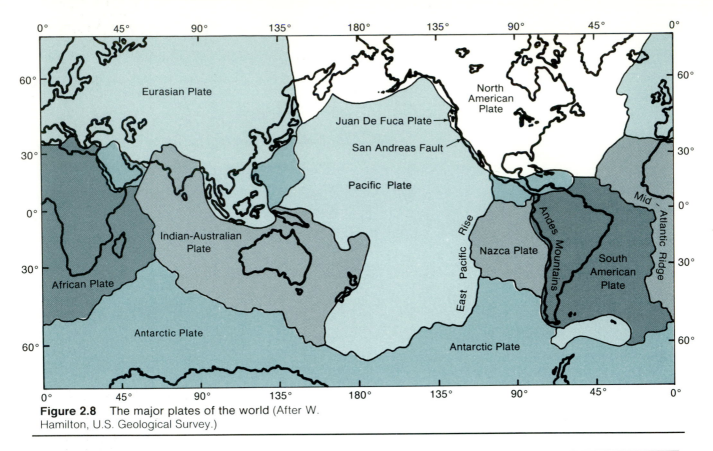

Figure 2.8   The major plates of the world (After W. Hamilton, U.S. Geological Survey.)

The lithosphere is not a continuous, unbroken structure but is, rather, composed of eight major plates and approximately twelve smaller ones (fig. 2.8). These plates are in motion, slide over the asthenosphere upon which they "float," and move toward, away from, or past each other at **plate boundaries.** The plates move apart at diverging boundaries, movement that causes tensional cracks—fissures—to develop in the crust. Magma from the mantle enters these fissures and solidifies. Since most diverging boundaries occur on the floor of the deep ocean, the solidified magma serves to form new oceanic crust.

Offshore Topography

The floor of the sea can be divided into the continental margin, which consists of the continental shelf and the continental slope, which are always present; a continental rise, which may or may not be present; and the deep-ocean basin, which is offshore both the continental slope and the rise, if present (fig. 2.9). The continental margin is composed of

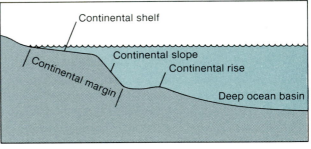

Figure 2.9   The sea floor. The continental slope marks the true edge of the continents. Both the shelf and the slope, though often covered by sediments, are composed of granite. The deep ocean basins are composed of basalt.

granitic continental crust, while the deep-ocean basins consist of basaltic oceanic crust. The continental slope marks the seaward extent of the continents; the continental shelf, the slope, and the overlying waters are considered to be the coastal ocean.

The rugged coastline of Southern California can be used to infer the offshore topography.

The continental shelf slopes gently seaward, with water depths of generally less than 130 meters (426 feet) and an average width of 65 kilometers (40 miles). The width varies with geographic region however; for example, the continental shelf of the east coast is wide, with a maximum width of 500 kilometers (311 miles) off Newfoundland, whereas offshore California the shelf is very narrow.

The topography of the continental shelf appears to mimic the topography of the adjacent landmasses. If the landmass has a gently sloping topography, as does the east coast of the United States, the topography of the shelf is also gently sloping. The topography of coastal southern California is rugged; therefore, the continental shelf is also rugged.

The crust of the continental shelf is generally covered by terrigenous sediments, sediments that are derived from the land and carried to the sea by erosional processes, river flow, and so forth. The coarser, heavier sediments, such as coarse and medium sand, are found closest to shore, where strong coastal currents prevent the finer materials from settling out. Farther offshore fine-grained mud overlies the continental crust. On narrow shelves the fine-grained mud may even extend over and cover all, or a portion of, the continental slope. On very wide shelves the outer portions may be covered with coarse sands that were presumably deposited near shore during glacial periods, when huge quantities of water were tied up as ice and sea levels were much lower.

The continental slope extends from a depth of between 100 to 200 meters (328 to 656 feet) at the edge of the continental shelf all the way down to the

Sediments are frequently drained off by submarine canyons and trenches.

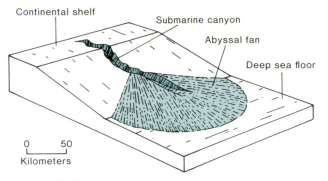

Figure 2.10 Sea fans and cones are formed by sediments moving down submarine canyons.

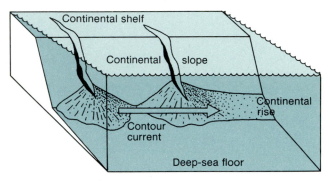

Figure 2.11 Sediments transported to the deep-sea floor are frequently moved about by contour currents. When these sediments are ultimately deposited, they form the continental rise.

deep-ocean basin. The slope is variable, with an average angle of between 4° to 5°. For example, offshore Baffin Island in the north Atlantic, it is smooth, gentle, and merges imperceptibly with the deep-ocean basin; off of Greenland the slope is steep and rather rugged.

Continental shelves and slopes are frequently intersected by submarine canyons. These canyons are generally V-shaped, with a width of several kilometers at the top and depths of up to 1,200 meters (4000 feet). Many canyons, such as the Astoria Canyon off the shore of the Columbia River in Oregon and the Hudson Canyon offshore the Hudson River in New York, are clearly related to and associated with rivers. Others, such as the Gulley, which is a canyon offshore Sable Island, Nova Scotia, are not associated with any rivers.

Sediment is transported down the canyons by seaward-moving bottom currents. If there is no trench immediately adjacent to the continental slope and canyon, the sediments will be deposited as sea fans and cones (fig. 2.10). These sediments, as well as those brought in from other canyons, are often moved along the base of the continental slope by contour currents (chapter 11), bottom currents that flow parallel to the base of the slope. The sediments are ultimately deposited to form the continental rise

(fig. 2.11), provided that there are no adjacent trenches to drain them off to even greater depths. Thus, the continental rise, where present, is composed of terrigenous sediments and overlies the basaltic oceanic crust.

The presence or absence of a continental rise indicates the geological activity in a given area. Since continental rises are generally found only in geologically inactive areas, those continental margins that have an associated continental rise, are often called **passive continental margins.** Conversely areas that lack a continental rise are frequently sites of geologic activity such as volcanoes and earthquakes. These continental margins are therefore termed **active continental margins.**

A passive continental margin consists of a continental shelf, slope, and rise. A broad, flat sediment-covered abyssal plain extends seaward beyond the continental rise. Active continental margins also have

 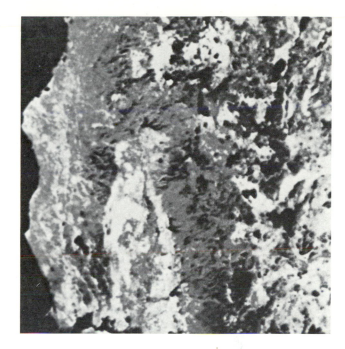

The abrupt bend in the east coast (a) and the west coast (b) may be due to fracture zones

a shelf and slope. The slope, however, ends abruptly at and forms the landward wall of an oceanic trench. Since the trench effectively drains off sediments, there is no continental rise associated with active continental margins. In addition the ocean floor seaward of the trench is irregular and hilly, rather than broad and flat as are the abyssal plains that characterize the deep-ocean floor immediately adjacent to passive continental margins.

Passive continental margins are found adjacent to landmasses in most of the Atlantic and Indian oceans and in the Pacific Ocean along the coast of the United States, where large sea fans are presently building off the coasts of California, Oregon, and Washington. These sea fans are coalescing to form a typical continental rise. Active continental margins are found throughout the remainder of the Pacific Ocean, as well as in areas of volcanic activity, such as offshore Iceland in the Atlantic Ocean.

Broad, flat, sediment-covered abyssal plains seaward of the continental slope, such as are characteristic of the topography directly offshore passive continental margins, are absent offshore active continental margins. Here the trenches are common immediately seaward of the continental slope and, beyond the trenches, the topography is rugged with

hills and depressions. It is thought that the deposition of terrigenous sediments has buried, and therefore obscured, the more rugged topography offshore passive continental margins, whereas the terrigenous sediments are trapped by the trenches found immediately seaward of active continental margins and, as a consequence, cannot move out onto and obscure the topography of the deep-ocean floor.

Farther offshore low abyssal hills rise from the ocean basin and reach heights of several hundred meters. Abyssal hills cover approximately 80 percent of the Pacific and 50 percent of the Atlantic Basin. They are, therefore, the most common topographic feature in the ocean basins. Many appear to be volcanoes, while others may have been formed by intrusions of molten rock that have pushed up the underlying sediment. Higher forms are termed seamounts, which appear to occur in groups. Seamounts and their associated forms with flattened tops, known as guyots, occur in large numbers in the Pacific Ocean, where they tend to form linear chains.

Fracture zones and faults also dominate the ocean basins. They are actually fractures in the oceanic crust, where one side of the sea floor has been displaced relative to the other side. In some areas these fracture zones offset the continental margins. For example, the abrupt bend in the coastline from Massachusetts to New Jersey on the east coast and that to the north of Los Angeles on the west coast have, most likely, been caused by fracture zones. These zones extend for thousands of miles throughout the ocean basins. In the Atlantic they are associated with the central ridge system, while in the Pacific they form chains of volcanoes and submarine mountains that are between 100 and 200 kilometers (62 and 124 miles) wide and consist of series of individual ridges and troughs that are hundreds of kilometers long and tens of kilometers wide.

In addition to the fracture zones, volcanoes and volcanic islands are also associated with "hot spots." In these instances molten lava pushes its way through the ocean's crust and eventually forms volcanic islands or seamounts. Such islands and seamounts are particularly conspicuous in the Pacific Ocean. Volcanoes generally form in groups. Some, like the Hawaiian Islands, form into long chains; others are circular. They generally rise several kilometers above the ocean basin, and as on land, are cone-shaped. These volcanoes tend to alternate between periods of activity and dormancy. Active periods may last for millions of years, during which the cone will build. Between eruptions the volcanoes generally subside as the crust and mantle undergo isostatic adjustment in response to the weight that was added during the eruptions. The presence of volcanic activity correlates with deep faults from which lava can move upward to the ocean bottom.

Ridges and offshore rises, not to be confused with the continental rise, occur in all of the world's major ocean basins, and all are intersected by fracture zones, major lines of weakness in the earth's crust. Fracture zones cross the ridges at right angles, and the sea floor on one side is often at a different height from the sea floor on the opposite side. Fracture zones are known to extend for thousands of miles across the deep-ocean floor and generally head directly toward the continental margins. It is possible that these fracture zones extend onto the continental landmasses themselves.

Most ridges and rises, such as the Mid-Atlantic Ridge and the Carlsberg Ridge in the Indian Ocean, are found near the center of the ocean basin. The East Pacific Rise, on the other hand, is closer to the South American continental margin. The Mid-Atlantic Ridge

The Hawaiian Islands have a volcanic origin.

Surtsey was formed in 1963 when a submarine volcano near Iceland erupted. Within a few months it rose above sea level to form a new island.

traverses the ocean basin in a north–south direction and effectively separates the North Atlantic bottom water (chapter 11) into an eastern and western water mass. This ridge rises from between 1 and 3 kilometers (.6 and 1.8 miles) above the ocean basin, has a width that ranges from 1,500 to 2,000 kilometers (932 to 1243 miles) and also has many fracture zones and volcanoes. The volcanoes in this area have formed such islands as Iceland, Surtsey, the Azores, and Tristan de Cunha.

Toward the central portion, the ridge is intersected by a steep-sided rift valley (fig. 2.12). This valley is 20 to 50 kilometers (12 to 31 miles) in width and 1 to 2 kilometers (.6 to 1.2 miles) deep.

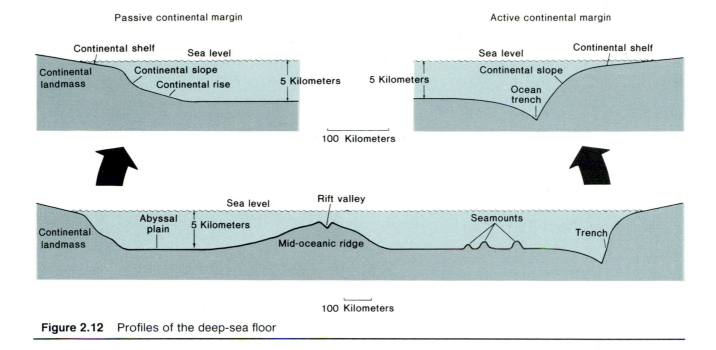

Figure 2.12 Profiles of the deep-sea floor

Rocks obtained from the rift valley floor are relatively young, having solidified from molten materials within the past few million years. The rift valley is bordered by steep-sided faults. These peaks rise to approximately 2 kilometers (1.2 miles) beneath the sea's surface. To the east and west of the central valley, the peaks are lower; the topography is, however, still quite rugged. Fracture zones and faults commonly intersect the ridge. One major fracture zone, the Romanche Trench, is a primary passageway through which water moves between the eastern and western portions of the Atlantic Basin. This trench is an example of a **transform fault,** which is defined as a fracture zone between two offset segments of an oceanic ridge crest.

The East Pacific Rise has a much gentler topography than the Mid-Atlantic Ridge. This rise is actually a low relief bulge in the ocean basin that gently ascends 2 to 4 kilometers (1.2 to 2.4 miles) above the basin and varies in width between 2,000 and 4,000 kilometers (1243 and 2486 miles). Despite its unspectacular topography, the East Pacific Rise is massive and is equal in size to both North and South America.

The East Pacific Rise travels northward from its location near South America and intersects North America in the Gulf of California. It continues northward beneath the continent and connects with the San Andreas Fault. The rise then appears off the shore of the Oregon coast and continues north to the Gulf of Alaska.

The ridge systems in the ocean basin are also areas of high heat flow from the earth's interior. These **ocean vents** appear to form in narrow bands, which indicate that they are narrow intrusions of molten rock entering the deep ocean. Recently a deep-ocean vent was found in the Galapagos Rift zone located between the Galapagos Islands and the mainland of South America. The water temperature in the vicinity of this vent was found to be between 8° and 16°C (46° and 61°F), which is considerably warmer than the normal oceanic bottom temperatures.

It is believed that the ridge systems provide the key to understanding the structure of the earth. As already noted the rocks collected in the rift valley of the Mid-Atlantic Ridge are only a few million years old, whereas those collected to the east and west of the ridge are older, with the oldest collected nearest to the continental slope. None, however, are older

Box 2.1 Life in the Rift Zone

The warm water entering the ocean in the Galapagos Rift zone contains high concentrations of various compounds of sulfur, which provide energy sources for bacteria. The bacteria utilize these compounds to produce materials that can be used as a food supply for several species of animal, most of whom were previously unknown to man—huge crabs, clams, worms, etc. Most of the animals found in these vent areas are much larger than their more common related forms. Interestingly, all of the animals were found within a 50-meter (164-foot) radius of these vents. Similar vents and their associated animals have since been found in the tropical Pacific Ocean and farther north offshore Vancouver Island, British Columbia. The presence of these animals is all the more remarkable because the deep ocean is so poor in animal life that it is generally considered to be a biological desert.

than 200 million years, which is much younger than the estimated age of the earth. These factors, along with other geologic evidence, indicate that the present configuration of the ocean basins and, indeed, the orientation and shape of the continents, are intimately related to occurrences at the oceanic ridge system.

The Sea Floor and Continental Wandering

The distribution of deep-ocean rocks of different ages is only one indication of the dynamic nature of the earth, its continents, and its ocean basins. Fossil coral, known to grow only in warm seas, is found in subpolar regions and suggests that these areas were once much warmer. Similarly coal deposits in Antarctica indicate that that continent was also much warmer since the organisms that compose these deposits required much warmer temperatures than those that

presently occur in this area. Furthermore the presence of identical, or at least very similar, rock and fossil assemblages found in such widely separated continents as South America, Africa, India, and Australia, provide evidence that these landmasses were much closer, if not joined, at some point in their history. In addition, the continental shelves of the continents appear to fit together, a circumstance that, when combined with the preceding observations, indicates that the continents were originally joined and then slowly moved apart to their present positions.

Perhaps the most convincing evidence is derived from paleomagnetism, which is the study of ancient magnetic fields. This involves the study of both igneous rocks, which are formed when magma cools, and sedimentary rocks, which are formed as sediments sink to the ocean floor and are compressed as a result of the tremendous pressure of the overlying water. Both types of rock contain small amounts of magnetic minerals.

As the molten igneous rock cools, the magnetic minerals "freeze" in a set position, giving the rock a measurable magnetic orientation. This orientation is established when the rock solidifies and does not change, even though the surrounding magnetic field may change drastically at some future time. The magnetic minerals in sedimentary rocks also align to the prevailing magnetic field as they settle and are compressed. This magnetic field, like that of igneous rock, is stable and does not change. Thus, both igneous rock brought to the surface as magma from the earth's interior and sedimentary rock from the earth's surface contain magnetic orientations that can be used to indicate the prevailing magnetic field of the earth when that particular rock was formed.

Studies of these materials illustrate that the earth's magnetic field has reversed itself many times, and that the frequency of the **magnetic reversals** has increased within the past 50 million years, averaging one reversal every 200,000 years. In other words it appears that the positions of the North and South poles have moved frequently in the past. This is known as polar wandering. The term *polar wandering* is, however, misleading since it suggests that the poles, themselves, have moved. In reality the magnetic reversals that are observed in rocks of different ages can result from one of two situations:

Box 2.2 Investigating the Sea Floor

The development of the echo sounder, or depth recorder, in the 1920s opened a new era in oceanographic research. This device measures the time that it takes for a sound pulse to leave a research vessel, travel to, strike, and reflect off the sea floor, and return to the ship. Thus, continuous measurements can be made while the vessel is underway.

The echo sounder was used in 1925 by the Meteor Expedition. Additional expeditions increased the data base and showed the rugged topography of the deep-sea floor.

The deep-sea drilling project was begun in 1968 and enabled oceanographers to retrieve rocks and sediments from the floor of the deep ocean. The data obtained from this project shows that the materials that compose the ocean floor are relatively young as compared with continental rocks, which are as old as four billion years.

Submersibles have provided much important information. For example, since 1977 the submersible ALVIN has descended to and explored deep-ocean vents. ALVIN is equipped with mechanical claws to enable it to obtain samples, as well as with cameras to allow it to obtain pictures and films of the sea floor.

The submersible ALVIN

either the continents remained stationary and the poles did wander, or the poles remained stationary and the continents moved in relation to the poles.

Paleomagnetic studies of rocks from a single continent could not resolve the question. However, when rocks of identical ages from different continents were analyzed, this comparative data conclusively illustrated that the continents, rather than the poles, had wandered. For example, the analysis of igneous rocks from North America shows that, if the pole did wander, it was located in central Asia some 250 million years ago. Analysis of identically aged rocks from Europe, however, shows that the magnetic orientation was closer to Japan during the same time. Thus, either there were two North poles, which is totally implausible, or the poles remained stationary and Europe and North America moved in relation to each other, as well as in relation to the stationary poles.

Consequently, *polar wandering* refers to the apparent motion of the poles. In reality what occurs is **continental wandering,** which can be traced directly to occurrences on the deep-ocean floor.

Box 2.3 The International Geophysical Year

Until the mid 1950s oceanographic studies were carried out by individual universities, research institutions, or countries. As the costs of mounting expeditions to study the deep sea and its floor continued to increase, it became obvious that an international effort was needed. As a result the International Council of Scientific Unions organized the International Geophysical Year of 1957–1958, and a team of scientists from 67 nations was formed.

During the International Geophysical Year, new data on heat flow from the earth's interior, as well as data on the earth's magnetism, was collected. When this data was compared and correlated with that which had been collected over the years, it indicated that the continents had changed their relative positions over geologic time. A theory of sea-floor spreading was then developed in 1962 in order to explain these phenomena.

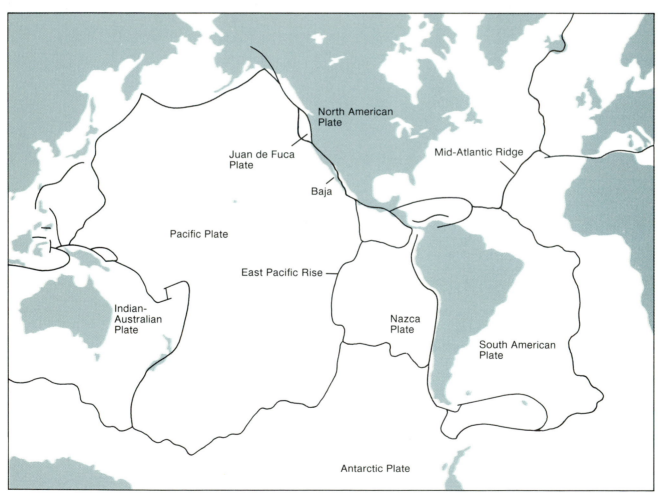

Figure 2.13 The major plates of North and South America. The majority of the United States is a part of the North American Plate. The west coast of the United States is, however, intersected by the East Pacific Rise. These plates are moving at different rates and in different directions. Since the Baja Peninsula and Los Angeles are located on the Pacific Plate, the Baja Peninsula will separate from the mainland and Los Angeles will move northward toward San Francisco, which is on the North American Plate.

Aerial view of the African Rift Valley

Continental Wandering and Plate Tectonics

Of the eight large and several smaller plates in the earth's crust and upper mantle (fig. 2.13), all the larger plates consist either entirely of oceanic crust and upper mantle or of a combination of oceanic and continental crust and upper mantle. The smaller plates may consist of oceanic crust and upper mantle; of a combination of continental crust, oceanic crust, and the associated upper mantle; or solely of continental crust and upper mantle. These plates may be termed **oceanic plates, continental plates,** or **mixed plates,** depending upon their characteristics. They "float" on the more fluid asthenosphere, and movement occurs at the plate boundaries. In general, all oceanic plates tend to move away from the oceanic ridges and rises and, as they move, may travel toward, away from, or past each other. As a result of the direction of movement, plate boundaries may be classified as **diverging boundaries,** which are boundaries between plates that are moving away from each other; **converging boundaries,** which occur when plates are moving toward each other; and **transform-plate boundaries,** where plates move horizontally past each other.

Since plates tend to move away from oceanic ridges and rises, diverging boundaries are common at these sites. Because the ridge crest is raised, the crust is thinner, and these areas are the site of volcanic activity. The ridge crest is bisected by a rift valley, which is formed as a result of these diverging plates. The divergence also forms tensional cracks, which are the sites where basaltic magma flows upward through the cracks and eventually solidifies to form new ocean floor. The new basaltic floor is added to the tailing edge of the plate. The result is that new ocean floor is formed at the rift valleys and is carried toward the continents by the moving plates. This tends to explain the age distribution of the rocks of the sea floor—the oldest rocks being found closest to the continental margins and the youngest closest to the ridges and rises.

When landmasses move apart at diverging continental boundaries, new seas are often formed. A diverging continental boundary, like its oceanic counterpart, is marked by a ridge or rise system where the divergence occurs. Since the crest is elevated, it is thinner at the raised, central portion, a characteristic that marks the area of volcanic activity where magma, presumably from the deeper portions of the mantle, rises to the surface. The African Rift Valley in eastern Africa exhibits all of these characteristics.

As this process continues, the continental crust will separate and, if it is near a sea-water body, sea-water will begin to enter the newly formed basin between the two diverging landmasses. This will lead

Box 2.5   Plate Movements and Topography

The movement of the Pacific Plate is quite rapid in comparison to that of the American and Atlantic plates. The different velocities, as well as the movements themselves, have had profound effects on the earth's surface and sea floor.

Before the San Andreas Fault was formed as a result of the movement of the Pacific Plate, the Baja Peninsula is believed to have been completely attached to the mainland of Mexico. Since the Baja is on the Pacific Plate, it moved—and continues to move—at a more rapid rate than the mainland, which is on the American Plate. The peninsula was formed when the Pacific Plate broke away from the mainland, creating the Gulf of California.

As noted previously, the East Pacific Rise has a gentler topography than the Mid-Atlantic Ridge. This is due to the different velocities of the Pacific

and Atlantic plates—the Pacific Plate is moving at approximately twice the speed of the Atlantic Plate, which is thought to result in the less spectacular topography of the East Pacific Rise.

Presently the Pacific Plate is moving north–northwestward at an average rate of 2 centimeters (about ¾ inch) per year. Los Angeles is on the Pacific Plate, while San Francisco is on the more slowly moving American Plate. A result is that Los Angeles is slowly moving toward San Francisco. Evidence of the rapid movement of the Pacific Plate can also be inferred from the rugged nature of the coastal mountains of California. The California coast ranges are considered to be young, developing areas that have been forming only during the last 65 million years as the Pacific Plate moves along the American Plate.

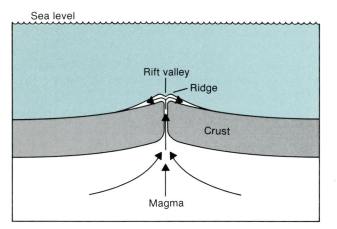

Figure 2.14 Magma comes to the surface of the deep sea floor and forms oceanic crust in the vicinity of submarine ridges.

to a pattern of faults and basins that will eventually become covered with terrigenous sediments derived from the newly separated landmasses. The edges of the diverging continents are thin, due to both the stretching and subsequent thinning of the crust at the divergence and the erosion of the continental edges. Erosion, as well as subsidence, will continue to lower the edges of the continents, which eventually will become flooded and will remain beneath the sea to form the continental shelves and slopes adjacent to the newly formed continents.

The central portion of the newly formed ocean basin will consist of a ridge bisected by a rift valley, which marks the original point of divergence (fig. 2.14). Basaltic magma from volcanic activity will continue to form new oceanic crust in the vicinity of the ridge. As the ocean continues to widen, the magma will cool and the edges of the ridge will subside and ultimately form the broad, flat ocean floor to either side of the ridge.

As oceanic and continental plates diverge, they will either approach other plates at converging boundaries or move horizontally past them at transform-plate boundaries. Faults commonly develop at transform-plate boundaries, particularly if the motion of the plates is not uniform. For example, the American Plate, upon which most of the United States lies, is moving west–northwest as a result of the divergence of the Mid-Atlantic Ridge. The Pacific Plate moves in a north–northwesterly direction at a more rapid rate, due to the divergence of the East Pacific Rise, which intersects North America in the Gulf of California.

These plates meet and move past each other at a more rapid rate than the American Plate. The well-known San Andreas Fault has developed at the plate boundary. The continual motion of these two plates, at different speeds, may, ultimately, cause the Baja Peninsula to separate from southern California and become an offshore desert island. Since much of the coastline to the north lies along this fault, it too may suffer a similar fate.

As noted, converging boundaries occur when two plates move toward each other. In general, three types of convergence are possible: oceanic plate–continental plate, oceanic plate–oceanic plate, and continental plate–continental plate. Should an oceanic plate converge with a continental plate, the denser basaltic oceanic plate will sink beneath the less dense, granitic continental plate. The sinking of a plate is called subduction, and the plate is said to be subducted. If two oceanic plates converge, the denser plate will be subducted beneath the less dense. In either case, subduction results in the formation of a trench at the converging boundary, and the subducted plate will form the outer wall of the trench.

As these plates continue to converge, the descending plate will come into contact with continually hotter material as it subducts into the mantle. The temperatures are high enough to cause the materials that compose the descending plate to partially melt. Since the resultant magma is less dense, it will rise toward the surface and erupt as a volcano. When the oceanic Juan de Fuca Plate (fig. 2.15) converged with the plate holding Oregon, Washington, and Vancouver Island, the magma came to the surface in the eruption of Mount Saint Helens.

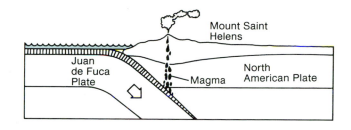

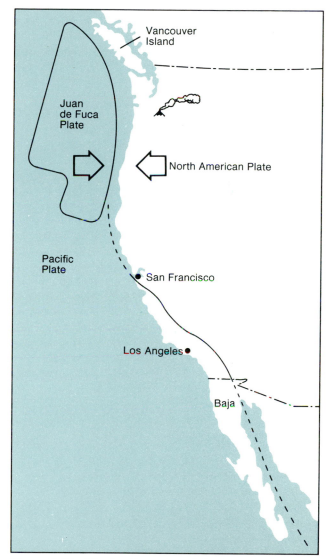

Figure 2.15 The leading edge of the Juan de Fuca Plate, composed of oceanic crust, converged with and was subducted beneath the North American Plate. Magma was formed and came to the earth's surface when Mount Saint Helens erupted.

The Sierra Nevada Mountains were formed when
an island arc collided with a continent.

Box 2.6 Table Mountain and the Earth's Mantle

Portions of the mantle and oceanic crust are exposed along the west coast of Newfoundland. At one time this portion of the continent was isostatically depressed and tectonic movements forced these materials over the continental margins. When the continent rebounded, the materials remained at the earth's surface. Mantle and oceanic crust form Table Mountain along the south shore of Bonne Bay. The lighter-colored mantle, composed of peridotites, overlies the oceanic crust.

The mantle peridotites produce toxic soil and water conditions. The result of these conditions is that the majority of Table Mountain is devoid of vegetation.

Table Mountain, Newfoundland

Where two oceanic plates converge offshore, the magma comes to the surface to form island arcs. Once formed, island arcs may converge with continents. In these cases the arcs are generally too light to be subducted and, as a result, will collide with the continent. Collisions of this type formed or added to the Sierra Nevada in California. As the plates continue to converge, the ocean floor seaward of the arc often breaks away and becomes subducted to form a trench offshore the original arc.

Should two continental plates converge, the continents will collide, but neither will be subducted; rather, a portion of the earth will rise as a result of the collision. The Himalayan Mountains were formed some 50 million years ago when Asia and India collided to form a single continent.

Mechanisms of Sea-Floor Spreading

Both convection and subduction have been suggested as the mechanisms that set plates in motion. Convection, which is the slow motion of a fluid or a molten substance away from the source of heat, causes magma, which becomes less dense as it becomes warmer, to rise up beneath the ridge crest (fig. 2.16). As it rises the magma could behave in one of two ways: it could either come to the surface at the rift valley and push the plates away from the center

and toward the continents (fig. 2.17), or it could flow laterally beneath the surface and pull or drag the plates toward the continents (fig. 2.18).

The presence of tensional cracks indicates that the plates are, indeed, pulled rather than pushed apart. Subduction of the leading edges of a plate into a trench would also tend to produce tensional cracks

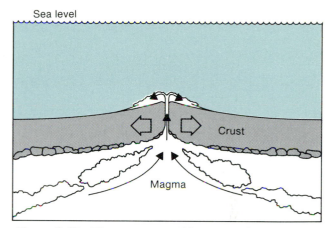

Figure 2.17 The magma could come to the surface of the sea floor and push the plates away from the central ridge.

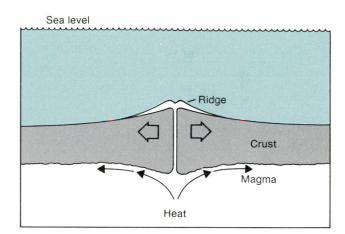

Figure 2.18 The magma may also flow beneath the oceanic crust, create friction, and pull the plates apart.

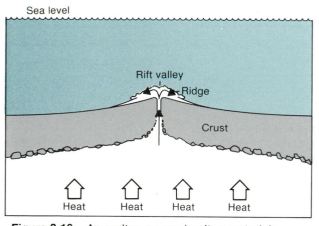

Figure 2.16 As molten or semimolten materials become warmer they will become less dense and move away from the source of the heat.

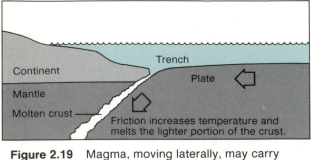

Figure 2.19 Magma, moving laterally, may carry the oceanic crust along with it. When the crust descends into the mantle, the lighter portions of the crust will melt. As this portion sinks through the mantle, it will pull the seaward portions of the plate along.

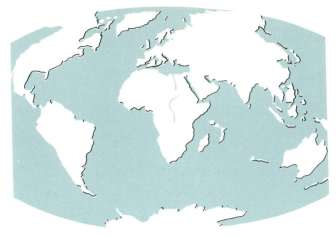

Figure 2.20 The present orientation of the continents

as the plates are pulled and sink. This process suggests that as the plate descends into the mantle, the lighter components of the crustal rock melt, leaving the heavier remainder on the leading edge of the plate. As a result the crust would rapidly sink into the mantle and pull the plate along (fig. 2.19).

Regardless of the mechanisms that cause plate motion, it is obvious that plates move in response to forces generated within the earth. These forces, known as tectonic forces, result in the uplift, movement, and deformation of the earth's crust.

The movement of the plates may be termed plate tectonics, a theory that explains many of the earth's phenomena. For example, plate tectonics explains the age distribution of rocks on the sea floor referred to on page 23. It is known that new oceanic crust is formed at the ridges and rises. Plate tectonics shows that this material will be slowly carried by the moving plate toward the continents and will ultimately be subducted at a trench. In the Pacific Ocean it is calculated that the sea floor is spreading at a rate of four centimeters (1.6 inches) per year. The maximum distance from the Pacific Rise to the farthest trench is 10,000 kilometers (6,215 miles). As a consequence newly formed oceanic crust in the Pacific Ocean should be subducted in approximately 250 million years.

Plate tectonics also serves to explain the different magnetic orientations observed in rocks of different ages. As the molten rocks were solidifying, the continents were in continually different positions in relation to the poles. With their magnetic orientations changing, the continents did, and continue to, wander in relation to each other and to the poles. Moreover the striking similarities in fossil and rock assemblages on widely separated continents is also easily explained by plate tectonics.

Continental Drift

The concepts of sea-floor spreading and plate tectonics explain the present structure of the ocean basins and the spatial orientation and shape of the continents (fig. 2.20). It is believed that approximately 250 million years ago there was only a single landmass called Pangaea (figs. 2.21 and 2.22). This continent began to separate about 180 million years ago. Initially the separation formed a long, narrow basin—similar, perhaps, to the Red Sea (fig. 2.23)—that was oriented in an east–west direction. Eventually this basin, known as the Sea of Tethys, formed a permanent waterway that separated Laurasia to the north from Gondwanaland to the south (fig. 2.24). Eventually Australia and Antarctica separated from Gondwanaland and moved south. This occurrence

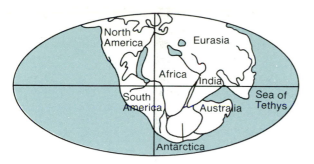

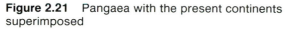

Figure 2.21 Pangaea with the present continents superimposed

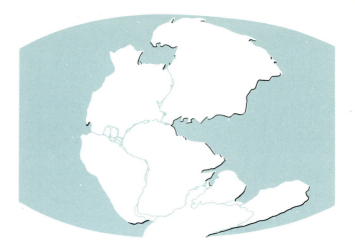

Figure 2.22 Pangaea 200 million years ago

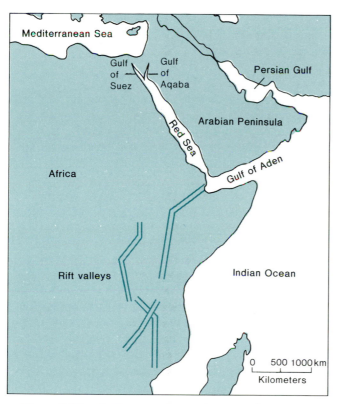

Figure 2.23 As Pangaea began to separate, a long, narrow basin was formed. This basin was probably similar to the present position of the Red Sea, which is an example of a diverging margin.

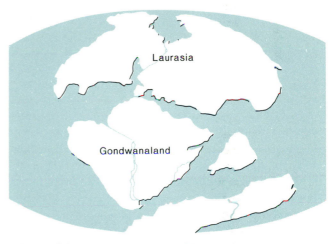

Figure 2.24 Approximately 100 million years ago, the basin had widened sufficiently to completely separate Pangaea. This circumstance formed Laurasia to the north and Gondwanaland to the south. The water body is called the Sea of Tethys.

Figure 2.25 Shortly thereafter South America separated from Africa.

was followed by the separation of North America, which took place approximately 100 million years ago and was followed shortly thereafter by the separation of South America (fig. 2.25). This separation was due to the formation of a north–south ridge system that began under Laurasia and continued southward beneath Gondwanaland. This process formed the Atlantic Ocean. Today the sea-floor spreading still continues: the Atlantic Ocean is still widening; the North American continental mass is moving northwestward along the San Andreas transform-fault boundary; and the Pacific is closing.

After the formation of North and South America, India separated from Africa, moved northward, and ultimately collided with the southern portion of Asia. The continents fused together along a dipping suture zone and the land buckled upward to form the Himalayan Mountain chain, the highest in the world.

The separation of a single landmass into the present continents would have also separated plant and animal populations. As the continents continued to drift apart, the ever-widening ocean would have presented an even greater barrier to prevent these populations from rejoining. As a result a single breeding population would become effectively separated. Since the separated populations could be expected to be exposed to very different environmental conditions, different traits, body forms, and other features would be accentuated. In time the individuals comprising the separated populations could be expected to become recognizably different. Ultimately these populations became so different that they became recognized as totally different species.

A more extreme case occurred when Antarctica separated from Gondwanaland. Initially the single land-mass was closer to the tropics. When it separated, the animals and plants were carried into a much colder environment. None were able to survive the polar cold; their remains form the Antarctic coal deposits.

Box 2.7 Mammals, Reptiles, and Continental Drift

Continental drift can be used to explain biological as well as geological phenomena. For example, biologists have long been puzzled over the fact that the Age of Reptiles lasted for 200 million years and resulted in only twenty major reptilian groups, while the Age of Mammals, which immediately followed, has lasted for only 65 million years yet has led to the development of thirty major groups.

Paleomagnetic studies have shown that during the Age of Reptiles there were just two supercontinents—Laurasia and Gondwanaland—separated by the Sea of Tethys. Initially the Sea of Tethys was narrow to its west and wider to the east. In addition there were land connections such as islands or true land bridges in the western portion of this sea. Because these connections enabled reptiles to move freely between Laurasia and Gondwanaland, the different populations were able to interbreed and there was little geographic isolation.

At the dawn of the Age of the Mammals, continental drift was well underway and the continents were widely separated. Also, sea levels were higher, the continental margins were flooded, and inland seas were formed. In some cases these seas completely separated the continents. For example, South America was completely bisected by seawater in the vicinity of the Amazon Basin. Under these conditions the mammals on each continent were isolated from those on the others, could not interbreed and, thus, rapidly diverged into several different groups.

Summary

The earth was formed as a result of the coalescence of diffuse clouds of dust and gas. As this material came together, the repeated collisions of the particles increased the temperature of the forming earth, and these materials became molten or semimolten. In that form the denser materials were able to sink, which resulted in a central core composed of iron, nickel, and sulfur, while the less dense basalts and granites rose and formed the surface crust. Since the basalts are somewhat denser than the granites, they were exposed as oceanic crust, while the less dense granite came to the earth's surface as continental crust.

The lithosphere consists of the crust, either oceanic or continental, as well as the upper mantle. This assemblage floats on the more fluid asthenosphere. The lithosphere is actually divided into plates, each plate moving about independently in response to either pushing by convection or—as seems more likely, due to the presence of tensional cracks—by pulling via convection or subduction.

The plates meet at converging boundaries, separate at diverging boundaries, or move past each other at transform-plate boundaries. When two continental plates diverge, they may form new ocean basins. When two oceanic plates diverge, they widen an existing ocean, as occurs in the Atlantic. These plates also converge with continental plates in the process of spreading apart, resulting in an oceanic plate–continental plate converging boundary. Since the oceanic plate is denser, it will be subducted beneath the continental plate to form an oceanic trench. As the oceanic plate subducts into the mantle, the temperature will rise and magma will form. This magma may then come to the surface, as was the case when Mount Saint Helens erupted. Should two continental plates converge, the land will buckle to form a mountain range and the landmasses will fuse at a suture zone.

When two oceanic plates converge, the denser will sink beneath the less dense, forming trenches, and magma will result and come to the surface to form island arcs. These island arcs may then converge with a continent to create a coastal mountain range such as the Sierra Nevada.

Plates may also move horizontally past each other at a transform-plate boundary. This occurs where the Pacific Plate moves past the American Plate. Since these two plates are moving at different speeds, a fault, such as the San Andreas Fault, will often form at the plate boundaries. As these two plates continue to move, the Baja will become separated from the mainland to become an offshore desert island.

The theory of plate tectonics explains many of the earth's phenomena. It accounts for the age distribution of sea-floor rocks, the magnetic reversals of both marine and terrestrial rocks, and the presence of similar or identical fossils and rocks on widely separated continents. It also provides the biologist with a model to aid in explaining the structure of present-day plant and animal populations.

Review

1. Compare and contrast active and passive continental margins.
2. Distinguish between converging, diverging, and transform boundaries.
3. When continental and oceanic plates meet at a converging boundary, which plate is generally subducted? Why?
4. Explain how magnetic reversals, fossil evidence, and the distribution of present-day organisms serve to explain the theory of continental wandering.
5. Explain how a single landmass could have separated to form the present day continents.
6. Compare and contrast the two mechanisms that may serve to set plates in motion.
7. How are sea arcs formed?
8. What happens when sea arcs collide with a continent?
9. What is the significance of the African Rift Valley?
10. What is the significance of the age distribution of the rocks found on the deep-ocean floor?

References

Burchfiel, B. C., Oliver, J. E., & Silver, L. T. (1980). *Studies in Geophysics: Continental Tectonics.* Washington, D.C.: National Academy of Sciences.

Dietz, R. S., & Holden, J. C. (1970). Reconstruction of Pangaea: Breakup and dispersion of continents, Permian to present. *Journal Geophysical Research,* LXXV, no. 26.

Glen, W. (1975). *Continental Drift and Plate Tectonics.* Columbus, Ohio: Charles E. Merrill.

Hallan, A. 1973. *A Revolution in the Earth Sciences: From Continental Drift to Plate Tectonics.* New York: Oxford U Press.

For Further Reading

Burchfiel, B. C. (1983, September). The continental crust. *Scientific American.*

Francheteau, J. (1983, September). The oceanic crust. *Scientific American.*

Kurten, B. (1969, March). Continental drift and evolution. *Scientific American.*

Plummer, C. C., & McGeary, D. (1979). *Physical Geology.* Dubuque, IA: Wm. C. Brown Publishers.

3
Sediments of the Shelves and Sea Floor

Key Terms

biogenic ooze
biogenous sediment
calcareous ooze
chemical weathering
cosmogenous sediment
foraminiferal ooze
hydrogenous sediment
lithogenous sediment
mechanical weathering

neritic sediment
ooze
pelagic sediment
poorly sorted sediment
radiolarian ooze
relict sediment
siliceous ooze
well-sorted sediment

A wide variety of sediments are found on the continental shelf and on the floor of the deep ocean. These sediments may originate from the land, the sea, animals, plants, and even from outer space. Regardless of their origin, all sediments are moved about and sorted by the water both prior to and after being deposited on the sea's bottom. Organisms living on or in the bottom sediments also rearrange and sort those sediments, while the types of the sediments influence the distribution of these bottom-dwelling marine organisms.

Marine Sediments

As a result of their different points of origin and their movement by the water, marine sediments are often distributed over the sea floor in particular patterns. These sediments are generally classified on the basis of their size, their origin, or their position on the sea floor in relation to the continents (table 3.1).

Sediment Size

Sediments are most often classified on the basis of their particle size (table 3.2), which determines the ease with which the sediments are moved about by the water and the rate at which they will settle to the bottom.

In general, the larger the particle, the greater its weight; therefore, water moving at a very slow velocity is able to transport only the very fine materials, whereas the larger-sized sediments, such as cobbles and pebbles, are moved about only by rapidly moving water. Similarly the rate at which particles settle out of suspension is a function of both their size and the velocity of the water in which they are suspended (fig. 3.1). Sand will tend to settle out rather rapidly; silt can be transported for considerable distances before it settles; and the finer clays will remain suspended indefinitely and will be carried for very long distances.

Table 3.1
Methods of Classification

Classification	Sediment Type	Source
Origin	Lithogenous	Rocks
	Biogenous	Animals and rocks
	Hydrogenous	Chemical reactions
	Cosmogenous	Outer space
Size	Boulders to clay	Primarily lithogenous
Location Shelf, slope, rise	Neritic sediment	Lithogenous, biogenous, hydrogenous, and cosmogenous
Deep-sea floor	Pelagic sediment	Lithogenous, biogenous, hydrogenous, and cosmogenous

Table 3.2
Size Classification of Sediments

Type	Size (millimeters)
Boulders	> 256
Cobbles	64–256
Pebbles	4–64
Granules	2–4
Sand	0.062–2
Silt	0.004–0.062
Clay	< 0.004

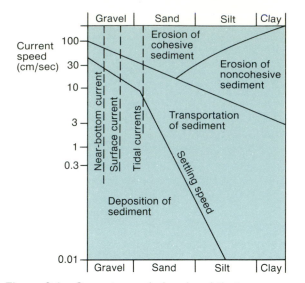

Figure 3.1 Current speeds (cm./sec.) that are required to erode, suspend, transport and settle sediments.

Well-sorted sediment

Poorly sorted sediment

A greater water velocity is, however, required to place clay particles into suspension than is required to keep them suspended. This is due to the particles' flatness, which provides them with a large surface area in relation to their weight and results in adjacent clay particles adhering to one another by attractive forces. Relatively high water velocities are required to overcome these attractive forces and separate clay into individual particles. Once the attractive forces are overcome, the individual particles will become suspended and remain so, even in very slow-moving water.

Sediments that consist of similarly sized particles are considered to be well sorted; the larger particles will have been deposited elsewhere and the smaller, lighter particles will have been removed by slower moving water. **Well-sorted sediments** have generally been subjected to sorting by waves and currents for a considerable period of time. Sediments found along ocean beaches are generally well sorted; the particles have been segregated by size as a result of both wind and wave action (chapters 10, 11).

Poorly sorted sediments, and unsorted sediments, consist of a wide variety of different-sized particles. These sediments have only recently become subjected to the actions of waves and currents. In northern coastal areas where glaciers meet the sea, the sediments, eroded from terrestrial rocks by glacial grinding, are generally poorly sorted and range from boulders to silts. As these sediments continue to remain in contact with the sea, they will be sorted by the waves and currents. The boulders will generally remain in place, while the finer materials will be transported and ultimately deposited elsewhere.

All of these sediments will continually be subjected to **mechanical weathering,** which will reduce their size. For example, even moderate wave action often causes pebbles to collide and break into smaller particles. Particles recently broken off from large rocks are poorly sorted, highly angular, and are generally found close to the rocks, pebbles, etc. from which they were formed. Sediments that have been exposed to the action of water for longer periods are better sorted and are said to have a low angularity, since they are rounder. Eventually these particles will become small enough to be carried by the water.

The predominant sediment size found in any given area can be used to indicate the magnitude of water motion. Along an open ocean beach, for example, the strong waves are responsible for depositing coarse sand and removing the finer sediments that eventually settle out in calmer, more protected areas such as bays and lagoons. When the water containing this material enters an inlet into the bay, the water will disperse and its ability to carry sediment—its carrying capacity—will decrease. The heavier material will be deposited in the vicinity of the inlet, while the finer materials will be carried into calmer waters before they, too, settle.

By analyzing the size and distribution of these sediments, it is possible to infer the velocity of the water in a given area. To illustrate, surface sediments that are composed of very coarse materials indicate that the water velocity is high, whereas silts would indicate very slow-moving water.

Not only can present conditions be determined, but it is also possible to infer past environmental conditions by examining sediment from the deeper underlying levels. Oceanographers commonly obtain sediment profiles by using a corer. A typical corer (fig. 3.2) has a sharp cutter head, which enables the core barrel to penetrate into the sediment. The core barrel contains a hollow core tube, which collects the sediment. Weights located above the barrel drive it deep into the sediment. When a corer is dropped overboard, the cutter head penetrates the sediment and the weights drive the barrel downward. The sediment is trapped in the core tube when a flap seats on the upper portion of the barrel and creates suction. The process is analogous to trapping liquid in a straw by holding one's finger over the upper end of the straw. The entire corer is then brought back on board, and the core tube containing an undisturbed sediment profile can be removed and the contents analyzed.

The sediment at the top of the core tube was deposited most recently, while that which is lower in the tube was deposited at progressively earlier times.

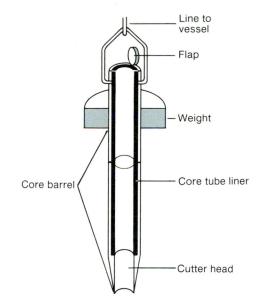

Figure 3.2 A typical coring device

A sediment profile

The sediment is generally removed from the core tube and divided into sections from top to bottom. Each section is commonly analyzed for its sediment size distribution.

In addition, the biogenic components of each section may be examined and the durable outer coverings of microscopic animals and plants identified. These materials are commonly deposited on and incorporated into the sediment after the organisms die and their remains sink to the bottom.

Box 3.1 Reconstructing the Past

The analysis of core samples is valuable, since it can be used to indicate environmental conditions that occurred far in the past. For example, if a core sample is taken from a protected bay and the sections are analyzed for sediment size, it may be found that the surface sediments, deposited most recently, are very fine and that the older sediments farther down the core tube become progressively coarser. This could indicate that the bay was open to the ocean at some time in the past and that water velocities were much greater. The progressive decrease in the size of the more recent sediments could then be correlated to the growth of sandbars and other such features (chapter 12) that could have provided protection from the open ocean and allowed the deposition of finer sediment.

The presence of the remains of small animals and plants in core sections can also be used to infer past conditions. Some of these living organisms are found only in cold waters, while others are known to require much warmer conditions. Finding the remains of cold-water animals and/or plants in sediment sections underlying seas that are presently much warmer would indicate that the water temperatures were colder at one time.

Sediment distribution

Hard and Soft Sea Floors

Sediment type is the major factor that controls the distribution of bottom-dwelling organisms, which are known as benthic organisms (chapter 8). The majority of these organisms have developed specific adaptations that enable them to live on or in particular sediment types. As a result these organisms have such specific requirements that it is frequently possible to infer the type of organism that will be found in a given area by a knowledge of the sediment (chapter 13). It is generally sufficient to classify the bottom as hard or soft for these purposes. Hard bottoms are generally found underlying areas where the water is moving rapidly and consist of sediment sizes that range from boulders to pebbles. Bottoms covered by shells or large shell fragments are also hard bottoms. Soft bottoms are composed of granules, sand, silt, and clay. The water above these bottoms is generally calm and moves with little velocity.

Sedentary organisms found on hard bottoms must be capable of either attaching onto or boring into the boulders, rocks, etc. The more mobile organisms, such as the common periwinkle and the starfish, must be able to move over the bottom while simultaneously attaching themselves firmly enough so that they are not swept away by the rapidly moving water. More sedentary organisms, such as the common barnacle, are able to cement themselves onto the bottom and are found on the surface of

Pacific sea star

Tube worm and burrow

boulders. Some types of worms and sponges actually bore into the rocks. The boring sponge, for instance, secretes chemical substances that dissolve the bottom sediments.

Organisms living on or in soft bottoms are often divided into three size classifications in order to readily relate them to the sediment sizes with which they are associated. Organisms whose shortest dimension is greater than 0.5 millimeters (greater than .02 inch) are termed the macrobenthos; those whose shortest dimension is between 0.5 millimeters (.02 inch) and 0.1 millimeters (.0004 inch) comprise the meiobenthos; and those whose shortest distance is less than 0.1 millimeters (.0004 inch) are named the microbenthos. Since soft bottoms are composed of fine sediments and are not subjected to strong waves, currents, etc., the organisms associated with these bottoms are not in danger of being swept away by waves, but they must cope with the rapid deposition of fine sediments. To do so, many of them tend to burrow into the bottom and have developed long tube-shaped body forms. The larger macrobenthos are able to push through the sediment, but the small

meiobenthos must move between the sand grains. To the smallest, the microbenthos, the fine sand is actually a hard bottom.

The life-styles of these organisms often serve to sort the sediments in which they are found. For example, the tube worm, found in waters off the east coast of the United States, burrows into soft bottoms, constructs a tube, and lives head down in this structure. This worm feeds by taking in sediments less than 1 millimeter (.04 inch) in diameter and utilizing food that clings to the surface. These sediments, along with wastes, are then excreted at the surface of the worm's tube. Thus, sediments are moved about and re-sorted by the tube worm's feeding behavior. Sorting of this type is called biogenic sorting.

The Origin of Sediments

Sediments may be classified on the basis of their origin as **lithogenous, biogenous, hydrogenous,** or **cosmogenous.**

Table 3.3 Transport of Terrigenous Sediment by the World's Major Rivers	River	Billion Tons Per Year
	Yellow, or Hwang Ho	2.1
	Ganges	1.6
	Amazon	0.4
	Mississippi	0.34
	Mekong	0.19
	Colorado	0.15

Bed load

Lithogenous Sediments

Lithogenous sediments are derived from rocks that are located either on land or beneath the sea. If the sediments are derived from terrestrial rocks, they are termed terrigenous sediments.

Regardless of their origin, all lithogenous sediment is the result of both the chemical and mechanical weathering of rocks. **Chemical weathering** is due to the tendency of certain materials to dissolve in water (chapter 7). These soluble materials go into solution and, when they reach the sea, contribute to the load of dissolved materials—salinity—that is characteristic of seawater (chapter 7). As the soluble portions dissolve, the rock becomes smaller until eventually only the insoluble portions remain. These smaller particles are more easily moved about by the water, which facilitates mechanical weathering. The particles are broken down into smaller and smaller fragments by repeated collisions with other rocks until they become small enough to become suspended and transported by the water. Chemical weathering generally dissolves soluble materials and places them in solution (chapter 7), while mechanical weathering physically reduces the insoluble particles to such an extent that they are able to be transported by the water. Ultimately both dissolved and suspended materials are carried to the sea.

Rivers transport approximately 20 billion tons of terrigenous sediment per year to the sea (table 3.3) and are by far the major sources of sediment for the world's oceans.

The finer sediments remain suspended within the river water, while the larger materials roll or slide along the river bottom or bed. This bottom material is known as the bed load of the river. Since the bed load consists of large, heavy materials, it is generally

deposited where the river water slows down when it meets incoming marine water. The finer suspended sediments are carried beyond this point and into the marine environment, where they may be transported along the coastline by various currents (chapters 4, 12). The heavier suspended material, on the other hand, will settle out close to the mouth when the river water loses velocity as it disperses into the sea.

The majority of this terrigenous sediment is derived from arid and semiarid regions (fig. 3.3) where rainfall is too low to support a vegetational cover capable of resisting erosion, yet is sufficient to cause considerable erosion. Tropical areas, on the other hand, supply little sediment to the world's sea, since these regions are extensively vegetated and therefore resistant to erosion. In polar regions large quantities of water are tied up as ice, which retards both chemical and mechanical weathering. In addition the tundra is dotted with glacial lakes that serve as sediment traps and prevent sediment from entering the sea. As a result polar regions supply little sediment to the world's oceans.

Wind erosion is also important in transporting sediment to the marine environment. It is estimated that winds remove approximately 100 million tons of terrigenous sediment per year. Most of this material is also removed from arid and semiarid regions of the earth. For example, a noticeable "plume" of sediment is generally present off Africa, where sand eroded from the Sahara Desert is transported for hundreds of miles. Since only relatively small, light

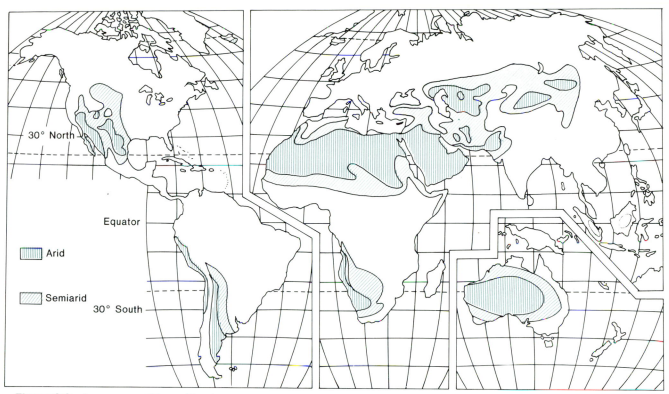

Figure 3.3 Large quantities of terrigenous sediments are derived from arid and semiarid regions. (From map by U.S. Department of Agriculture.)

Arid

Semiarid

30° North

Equator

30° South

Large quantities of terrigenous sediment are removed from arid regions during wind storms. This material is easily eroded due to the paucity of vegetation in these areas. Note the blowing sand, much of which will ultimately enter the sea.

materials are capable of being picked up and transported by air currents, much of this material is carried far out to sea and is ultimately deposited in the deep ocean. In addition very fine material, such as volcanic ash, may be carried completely around the world by upper air currents.

Biogenous Sediments

Biogenous sediments are derived from animal and plant materials. The most common of these sediments found on the sea floor are the protective coverings produced by microscopic and semimicroscopic marine animals and plants (chapter 8). Protective coverings from groups of microscopic plants, termed diatoms and silicoflagellates, are the most common plant materials found on the sea floor. The protective coverings from the single-celled foraminifera and radiolarians are the most common animal materials found in biogenous sediments. The silicoflagellates, diatoms, and radiolarians have protective coverings, consisting of compounds that contain the element silicon; the protective coverings of the foraminifera generally consist of compounds containing calcium. These materials, originally dissolved in the water, are removed by the organisms and used to produce their protective "shells," termed tests, which ultimately become deposited on the sea floor when these animals and plants die. Although the organisms themselves decompose, the durable protective coverings persist and become a part of the sea-floor sediment. Since many species of foraminifera are extremely sensitive to environmental change, the presence of these microfossils provides a record of past conditions in that particular area. Shells of small, floating marine snails, termed pteropods, are also found in localized areas, primarily in the shallow waters of the Atlantic Ocean.

When the biogenous material comprises more than 30 percent of the sediment, it is termed an **ooze.** Should radiolarians be the most common constituent, it would be called a **radiolarian ooze;** if foraminifera are the common constituent, it would be called a **foraminiferal ooze;** and so forth. These oozes may also be named for the most common element in the sediment. Thus, they may be termed **calcareous oozes** if the dominant element is calcium; **siliceous oozes** if the element silicon dominates, etc.

Excretory materials released by marine organisms also contribute to the biogenous sediments of the sea floor. Many marine organisms filter their food from the water column; others extract their food from materials that either live on or in the sediment or have been deposited on it. Whatever the case, unusable materials are taken in along with the food. These materials, as well as the waste products, are compacted and released as fecal pellets by these organisms. The fecal pellets ultimately become a part of the biogenic component of the sediment. Fecal-pellet-type materials produced by bacteria, as well as true fecal pellets, most likely supply a significant portion of the food for the bottom dwellers of the deep ocean (chapter 14).

Formation of biogenous sediment is controlled by three basic factors: the presence of the living organisms whose remains contribute to the biogenic sediment; the durability of the material after it is deposited on the sea floor; and the amount of nonbiogenic sediment that may serve to obscure the biogenic material.

Obviously the organisms that produce the outer coverings that predominate in biogenous sediment must be present in order for this material to be deposited. The microscopic plants, such as the diatoms and silicoflagellates, require various plant nutrients, as well as the element silicon (chapters 8, 9). Organisms such as the foraminifera, pteropods, and radiolarians generally feed on the microscopic plants. The presence of plant nutrients dissolved in the water column, as well as sufficient sunlight and the appropriate water temperatures, governs the distribution of the plants. The presence of plants, in turn, governs the presence or absence of the animals. In the North Pacific and in the waters adjacent to Antarctica, plant nutrients and silica concentrations are high, with the result that diatom oozes are common sediments in these areas. Diatom oozes are rarely found in the North Atlantic, however, due to both the low availability of the plant foods and the predominance of lithogenous sediments, which tend to obscure and/or dilute the diatom oozes.

In many instances the remains of these animals and plants tend to dissolve readily in the deeper water. This is particularly true of the calcareous materials that compose the shells of foraminifera and pteropods. Therefore, pteropod shells are found only on the tops of sea mounts and guyots in the Atlantic Ocean, while foraminiferal oozes are common only on the sea floors of the shallow oceans. Coral reefs (chapter 5) are biogenous sediments built by a specific group of marine animals that secrete an external skeleton of calcium in the form of calcium carbonate in seas where the water temperature is sufficiently high.

Lithogenous sediments often cover, and thereby obscure, biogenic material. This is most common in the vicinity of landmasses, where terrigenous sediment is carried into the marine environment by river flow and surface runoff. Biogenous sediments are also often covered farther offshore on the continental slope by lithogenous materials that are deposited by bottom currents.

Hydrogenous Sediments

Hydrogenous sediments are formed as a result of chemical reactions that occur between materials dissolved in the seawater itself. These reactions are apparently facilitated by the presence of dissolved oxygen, which is generally abundant in deep oceanic water (chapters 11, 14). The chemical reactions cause the dissolved materials to coalesce and eventually precipitate out of the water column. Once precipitated, the small particles are carried by currents until they encounter and adhere to solid surfaces such as rock fragments and small pieces of bone. Additional particles accumulate in the same manner, ultimately forming a large nodule. These nodules contain high concentrations of iron, manganese, copper, cobalt, and nickel.

The aforementioned elements are thought to be derived from one of three possible sources: they may have been carried to the sea dissolved in river water, surface runoff, or groundwater, in which case they would have a terrestrial origin. Another possible source could be the chemical weathering of newly formed oceanic crust (chapter 2). Or they could have been dissolved in the hot water that enters the ocean bottom in rift and vent areas (chapter 2); this water is commonly referred to as juvenile water (chapter 7).

Nodules are unable to form in areas where there is a large input of other sediments, since they tend to cover the forming nodule and thereby prevent the accretion of additional material. Consequently, and with the exception of the Blake Plateau (fig. 3.4), nodules are rare in the Atlantic Ocean, where there is a large input of terrigenous sediment. The Blake

Manganese nodules on an abyssal plain

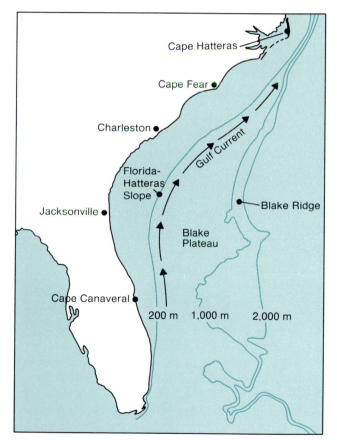
Figure 3.4 The Gulf Current moves directly over the Blake Plateau and prevents the deposition of sediments; this enables nodules to form.

Plateau is immediately offshore the coast of Florida, Georgia, and the Carolinas. The Gulf Current (chapter 10) moves directly over the bottom in this area and, presumably, prevents the deposition of lithogenous sediment. Because of the presence of trenches at the base of the continental margins, which act as traps for terrigenous sediments, sediments accumulate very slowly in the central portions of the Pacific Ocean. As a result nodules, particularly those composed primarily of manganese, are so numerous that they are estimated to cover over 35 percent of the Pacific Ocean bottom.

Cosmogenous Sediments

Cosmogenous sediments are derived from outer space and are formed by collisions between meteoroids and/or asteroids. As a result of these collisions, this extraterrestrial material is reduced to and falls to earth as dust that ranges in size from 200 to 400 microns (.008 to .016 inches). This material consists primarily of iron and iron-rich materials.

Sediment Distribution

The characteristics of the sediments that cover the sea floor change as the distance from landmasses increases. As a result sediments may be divided into **neritic sediments** and **pelagic sediments.** A third category, **relict sediments,** is used to distinguish atypical sediments that are found on the outer portions of the continental shelf.

Neritic sediments are composed primarily of terrigenous materials that are carried into the sea by rivers and surface runoff. Biogenous, hydrogenous, and cosmogenous materials are also found in neritic sediments. However, since terrigenous sediments accumulate at a more rapid rate, they tend to bury, dilute, and obscure the other materials. Neritic sediment is distributed along coastlines and is carried farther out on the continental shelf by coastal currents. In addition these materials move down submarine canyons to the deep-ocean floor where, if there are no trenches present, they may then be transported by contour currents (chapters 2, 11). Neritic sediments display a broad range of sizes. The larger, heavier materials tend to be deposited closer to shore, whereas the lighter, finer sediments are found farther offshore.

Box 3.2 Hydrogenous Sediment and Life in the Sea

In many areas, groundwater and river water contain large amounts of dissolved iron. Since these waters are acidic, the iron remains in solution. Seawater is, however, much less acidic and when the iron-rich ground and river waters meet the coastal sea, the iron begins to precipitate due to the low acidity. Initially the iron forms a surface film, often mistaken for an oil slick. As more and more iron coalesces, nodules form and react with dissolved phosphate—an essential plant nutrient (chapter 9). When this occurs, the phosphate goes out of solution and settles to the sea floor.

In areas where large quantities of iron-rich ground or surface water enter the coastal seas, significant amounts of phosphorous may be removed from solution. This upsets the nutrient balance and can adversely affect the sea's plant and animal life.

Pelagic sediment is found offshore the continental shelf on the floor of the deep ocean. These materials are generally very fine and accumulate at a much slower rate than the neritic sediments on the continental shelf. This is due to the distance from landmasses and river mouths, which are the major sources of terrigenous sediments. As a result and, although lithogenous sediment is a major component of pelagic sediment, hydrogenous and biogenous materials are not greatly diluted and obscured and are therefore very abundant on the deep-sea floor.

Relict sediments are composed of materials that were deposited during the ice ages when large quantities of water were tied up as glacial ice. During these periods sea level was much lower, and large portions of the continental shelf were above water. Consequently rivers met the sea much farther offshore, and the deposition of large, heavy terrigenous sediment occurred in the vicinity of these prehistoric river mouths. When the glaciers melted and the sea level rose, the relict sediments remained in place and now mark the original point of deposition. These sediments may be thought of as being out of equilibrium with their present environment, since they are found in atypical locations.

As noted earlier, neritic sediments are the major materials to be found overlying the oceanic crust and the continental shelf. The abundance and distribution of these sediments on the continental shelf are determined, primarily, by the topography of the coastal landmasses. For example, the Atlantic coast is protected by a series of barrier islands extending from Long Island, New York, to Georgia. Rivers empty into the protected bays located between these islands and the mainland, and as a consequence much of the terrigenous sediment brought in by river flow and surface runoff from the mainland is trapped in and deposited on the bay bottoms, rather than being transported farther offshore onto the outer portions of the continental shelf. Protected areas such as bays, which tend to trap sediments, are often referred to as sediment sinks. The result of this trapping is that the dominant lithogenous sediments on the outer portions of the continental shelf of the Atlantic Ocean are of two kinds: the finer materials that do not settle out in the calm bay water and the relict sediments.

Barrier islands are, for the most part, absent on the Pacific coast, where the rivers empty directly into the ocean. With the absence of protected bays, the sediments are carried farther out onto the continental shelf. Moreover the continental shelf of the west coast is intersected by numerous submarine canyons. These canyons drain off this sediment and carry it to the base of the continental slope, where in many areas it is again drained off into the trenches—the sites where the oceanic plate is being subducted (chapter 2). Although both the coarse and the finer neritic sediments are carried farther out onto the continental shelf in the Pacific Ocean, these sediments are prevented from traveling out onto the deep-ocean floor by the presence of the trenches at the base of the continental slope. The exceptions, as noted in chapter 2, are the passive continental margins, which are prevalent off the coasts of Washington, Oregon, and California. In these areas sea fans tend to coalesce to form a continental rise, and this sediment is then moved about by deep-sea currents.

Barrier islands provide protection to the mainland. The protected bays behind the barrier islands trap large amounts of sediment.

Although most of the neritic sediment is deposited on protected bay bottoms in the Atlantic and is drained off into submarine canyons and oceanic trenches in the Pacific, the very fine neritic materials are transported far offshore by winds or water currents and eventually do settle out on the floor of the deep ocean. These particles mix with biogenous, hydrogenous, and cosmogenous materials, as well as volcanic ash, to form the pelagic sediments of the deep ocean. Since all these materials consist of very fine, clay-sized sediments, they are often referred to as abyssal clays. Abyssal clays are reddish brown because of the presence of iron compounds.

Biogenic oozes are often the dominant materials in pelagic sediments (table 3.4). Their distribution is determined by the presence of the living organisms in the overlying water column and/or in or on the sediments themselves, as well as by the characteristics of the water that is immediately above the bottom sediments.

Deep waters are generally low in the element silicon, with the result that siliceous oozes are rare on the deep-ocean bottom. Also, silicon dissolves readily. Since the protective outer coverings of the silicoflagellates, diatoms, and radiolarians contain silicon, their outer coverings dissolve and the ooze

Table 3.4 Sediment Types in the World's Oceans	Ocean	Mean Depth	% Ooze	% Abyssal Clay
	Atlantic	3296 m.	76	24
	Pacific	4282 m.	51	49
	Indian	3963 m.	74	26

does not form. Similarly at great depths the concentration of carbon dioxide increases and readily converts to carbonic acid. As a result deep water, particularly at depths greater than 4,000 meters (13,123 feet), is generally acidic, and calcareous materials dissolve rather than form oozes below this depth. Table 3.4 indicates that the Pacific Ocean is the deepest of the world's oceans; consequently, it contains the lowest percentage of calcareous and siliceous oozes. In shallower portions of the deep ocean, diatomaceous oozes are common in the colder areas, while radiolarian oozes are found in the equatorial regions. These conditions reflect both the high concentrations of dissolved silicon and the dominant populations of diatoms and radiolarians present in the overlying water columns. Foraminiferal oozes are also common in the shallower Atlantic and cover more than 50 percent of the ocean floor.

The deposition of sediment onto the floor of the deep ocean is a very slow process. Commonly, less than 1 centimeter (.4 inch) of sediment is accumulated in 1,000 years. The sediment that covers the oceanic crust offshore the continental shelf is therefore very thin, having an average depth of only 0.6 kilometers (.4 mile). Pelagic sediments are thinnest in the vicinity of the ridges and rises. In some areas they are totally absent and the oceanic crust is exposed. There are two plausible reasons for the paucity of sediments in these areas: the ridges are newer features on the sea floor (chapter 2) and strong bottom currents keep these areas swept clean of sediments. The pelagic sediment becomes thicker away from the ridges and rises and is generally at its deepest over the older oceanic crust nearer the continental slope.

Summary

The sediments of the sea floor may be divided into lithogenous, hydrogenous, biogenous, and cosmogenous sediments. Lithogenous sediments, the major sediments on the ocean bottom, are derived from the chemical and mechanical weathering of rocks. If the lithogenous material is derived from the land, it is termed terrigenous sediment. This sediment is generally carried to the sea by river flow, surface runoff, or by winds.

Rivers are the major source of terrigenous sediments. When these materials enter the marine environment they are distributed and sorted by currents. Winds carry significant quantities of fine sediments far offshore, where they are eventually deposited to become major components of the abyssal clays.

Biogenous sediments are composed primarily of the protective outer coverings of small marine animals and plants. If these remains comprise at least 30 percent of the sediment in a given area, the sediment is termed an ooze. Oozes may be named for the types of organisms that formed them; for example, diatomaceous oozes and foraminiferal oozes. They may also be named for the major chemical elements present in each; for example, a siliceous ooze or a calcareous ooze.

Hydrogenous sediments form as a result of the chemical reactions that occur in seawater. These reactions ultimately result in the formation of small particles, which are deposited on the sea floor. Currents move these particles about and cause them to collide with other particles. Repeated collisions may form the nodules that are found on some portions of the deep-sea floor.

Sediment size and the remains of specific organisms in biogenic sediments are often used to infer past conditions in a specific area. Sediment-size analysis of core sections can be used to infer the speed of water, while the protective outer coverings of the organisms may be used to deduce other environmental conditions, such as water temperature.

In addition the sediment type frequently determines the type of organisms that will be found in a particular area. Hard bottoms will generally contain assemblages of burrowing organisms, as well as those capable of cementing themselves onto the rocks. Soft bottoms contain different types of burrowing animals, which commonly rework and re-sort the sediment.

Review

1. What are the three methods of classifying marine sediments? Explain each.
2. Distinguish between well- and poorly sorted sediments.
3. Explain how sediment size is used to infer the degree of water motion.
4. What are relict sediments? Explain their significance.
5. How can past conditions be determined by analyzing the sediments collected by coring?
6. Why and how does the sediment type determine the sort of organisms found on the sea floor?
7. What factors determine the presence of biogenic sediments on the sea floor?
8. How are hydrogenous sediments formed?
9. Compare and contrast nodule formation in the Atlantic and Pacific oceans.
10. Compare and contrast the distribution of lithogenous sediments on the sea floor of the Atlantic and Pacific oceans.

References

Keen, M. J. (1968). *An Introduction to Marine Geology*. New York: Pergamon.
Menard, H. W. (1964). *Marine Geology of the Pacific*. New York: McGraw-Hill.
Shepard, F. P. (1973). *Submarine Geology*. New York: Harper & Row.
Turekian, K. K. (1976). *Oceans*. New York: Prentice-Hall.

For Further Reading

Emery, K. O. (1969, September). The continental shelves. *Scientific American*.
Erickson, D. B., & Wollin, G. (1962, July). Micropaleontology. *Scientific American*.

4
Waves and Tides

Key Terms

apogee out-of-phase tide
centrifugal bulge
deep-water wave
diffraction
diurnal tide
elliptical orbit
gravity wave
lunar bulge
mass transport
mixed tide
neap tide
orbital motion
perigee in-phase tide
reflection
refraction
reversing tidal current
rotary current

seiche
semidiurnal tide
shallow-water wave
solar bulge
spring tide
surf zone
swash zone
tidal range
tilted elliptical configuration
tsunami
wave crest
wave form
wave height
wave length
wave period
wave train
wave trough

Waves are variable and transitory features of the sea's surface. They can range in size from the smallest ripple to the huge walls of water that are produced by disturbances of the oceanic crust. All waves are generated by one of three basic mechanisms: wind, gravity, or earthquakes.

Once waves are formed, they are capable of moving for long distances throughout the sea with little loss of energy. As they travel over the deep-ocean floor, most waves have little interaction with the bottom. When they move onto the rising continental shelf, and particularly when they enter shallow coastal areas, waves do begin to interact with the bottom. The result is a change in their shape, speed, and motion. In shallow water the waves will eventually become so modified that they will break on a shoreline and release a considerable amount of energy. Waves may also be deflected, refracted, and reflected by piers, islands, and other such things.

The actual shape of a coastline will affect the characteristics of waves. A classic example is the enhancement of tides in the Bay of Fundy. In this area the coastline of Maine, New Brunswick, and Nova Scotia alters the shape of these waters to create a large difference between high and low tide.

Wave Observations

The surface of the sea presents a confusing pattern of waves of different sizes and shapes, moving at different speeds and traveling in several different directions. To attempt to follow the progress of a particular wave or series of waves for even a short period of time is virtually impossible. Oceanographers generally begin the study of waves under controlled conditions. They do this by retreating into the laboratory, where they make and observe waves in a wave tank or wave channel.

A wave channel (fig. 4.1) is merely a long tank with glass sides. Waves are generated by a motor-driven paddle at one end. An artificial beach or shock absorber at the other end prevents the water from traveling back up the channel and producing confusing waves. The beach may become part of the actual experiment if one wishes to observe breaking waves.

Waves frequently present a confusing picture to the observer.

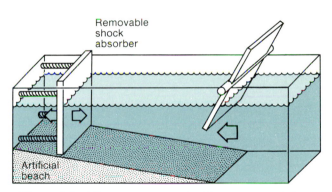

Figure 4.1 Waves are generated in a wave tank by a motor-driven paddle. A shock absorber at the other end of the tank prevents the waves from moving back up the tank. The shock absorber may be removed and sand added to form an artificial beach, should one desire to observe breaking waves.

The actual motion of the water in the wave tank may be traced by placing a marker at different depths. The marker, often referred to as gunk, is a mixture of light oil and zinc oxide, which has a white color. By varying the mixture of oil and zinc oxide, the density of the particles of gunk can be adjusted to that of the water in the tank and so can be suspended in the water at different depths. The gunk will behave identically to the water that it displaces; therefore, the movement of the water at different levels in the wave tank can be inferred from the movement of the gunk. Since the tank has glass sides, it is possible to trace the path of the gunk with a grease pencil, marker, etc., and obtain a record of the water movement.

By generating and observing waves in a wave channel, all the variables that confuse matters in the sea can be eliminated and waves closely approximating the ideal concept can be formed and studied. These waves, called sine waves (fig. 4.2), are actually representations of a mathematical expression for a smooth, regular oscillation.

Characteristics of Waves

Waves—whether traveling through the deep ocean, shallow coastal waters, or in a wave channel—all share common characteristics (fig. 4.3). All waves have a **crest** and a **trough**, as well as a specific **wave height, length,** and **period.** The crest is considered to be the high point of a wave; the trough is the low point. The wave height is the vertical distance from the crest to the trough; the wave length is the horizontal distance between adjacent crests. A series of waves traveling from the same direction is termed a **wave train.**

Waves are classified according to their wave period—the amount of time that it takes one wave crest to travel the distance of one wave length. Wave periods are generally measured in seconds; however, the wave periods for large waves, such as those generated by earthquakes, submarine eruptions, and the like, have wave periods that are measured in minutes. These huge, destructive waves are often erroneously called tidal waves by the general public. In reality most coastal areas are struck by tidal waves—gravity-induced waves—every day during periods of

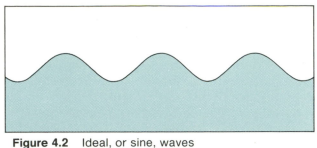

Figure 4.2   Ideal, or sine, waves

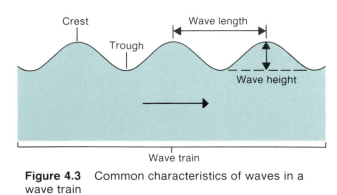

Figure 4.3   Common characteristics of waves in a wave train

high tide. True tidal waves are formed as a result of the gravitational attraction of the moon and the sun on the earth's waters and have periods of approximately 12 or 24 hours, depending upon the location of the landmass. In an effort to place true tidal waves in their proper perspective, oceanographers have adopted the Japanese word **tsunami** to describe waves generated by earthquakes. It was later found that *tsunami* means "tidal wave" in Japanese! Most likely Japanese oceanographers are now trying to find another word to describe these unique waves. Table 4.1 summarizes typical wave periods.

Three other measurements—wave frequency, speed, and steepness—can be easily made, particularly in a wave channel. Wave frequency is defined as the number of waves that pass a fixed point in a given period of time. Wave speed is the relationship of the wave length to the wave period. Wave speed can be readily calculated from the equation $C = L/T$, where C is the speed, L equals the wave length in meters, and T stands for the wave period in seconds. Wave steepness is the ratio of wave length to wave height. It has been experimentally determined that there is a definite relationship between the length,

Table 4.1	Wave Type	Period
Typical Wave Periods	Ripples	1 second
	Wind chop	1–4 seconds
	Fully developed seas	5–12 seconds
	Swell	6–16 seconds
	Surf	1–3 minutes
	Tsunamis	10–20 minutes
	Tides	12 or 24 hours

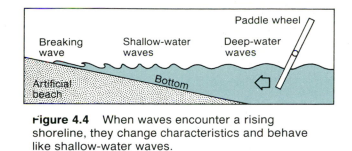

Figure 4.4 When waves encounter a rising shoreline, they change characteristics and behave like shallow-water waves.

height, and stability of a wave. When the wave steepness exceeds 1:7, waves become unstable and tend to break. In other words a wave that is 7 meters long can be no higher than 1 meter and still maintain its stability.

If a series of waves are generated in a wave channel with a rising shoreline (fig. 4.4), they will initially travel as deep-water waves; then when they encounter the rising bottom, they will become shallow-water waves. As this wave train is moving in deep water, it will be found that each wave moves at a speed that corresponds to its own particular wave length and period. If a specific wave is singled out and followed, it will be found that this wave will advance through the wave train and rapidly overtake previously formed waves. As it continues to progress through the wave train, it will gradually lose energy and decrease in height. When it reaches the front, it will disappear and be replaced by other more recently formed waves that have moved up from the rear. The disappearance of the leading wave is due to its loss of energy as it moves into undisturbed water. The lead wave must expend energy in order to set the undisturbed water in motion, and it is this loss of energy that causes the leading wave to die out. The fact that this invariably occurs indicates that each particular wave in a train actually moves faster than the wave group. Repeated observations indicate that in deep water the wave energy of the group is one-half the energy of an individual wave.

As the wave group moves into shallow water at the far end of the wave channel the waves interact with and are slowed by the bottom, until the wave speed of each equals the group speed. Actually the waves at the front of the wave group are the first to encounter the rising bottom and lose speed. Since the waves farther offshore are still in deep water, their speeds have not yet decreased and, as a result, the distance between the waves decreases. In other words the wave length shortens, which serves to destabilize the leading waves. The destabilization is intensified by the shallow bottom's tendency to force water from the trough into the crests of the leading waves, thus increasing their wave height. As wave length decreases, wave height increases; this upsets the 1:7 ratio of height to length, and the waves begin to break.

Since waves behave differently in deep and shallow water, they are often classified as **deep-water waves** and **shallow-water waves**. A wave is considered to be a deep-water wave if the ratio of water depth to wave length is greater than 1:2; it is classified as a shallow-water wave if the ratio of depth to length is less than 1:25. If the ratio of water depth to wave length is between 1:2 and 1:25, the wave is classified as an intermediate wave. This classification depends on the depth of the water, as well as on the wave length. In the same water depth, then, a wave of one length may be classified as a deep-water wave while another, with a longer wave length, may be classified as a shallow-water wave. For example, in 3 meters (10 feet) of water, a wave with a length of less than 1 meter (3 feet) would be considered to be a deep-water wave; at the same depth a tsunami with a wave length of 75 kilometers (47 miles) would behave as a shallow-water wave.

With the exception of the smallest ripples, all waves are termed **gravity waves,** since once crests and troughs are formed, gravity pulls the waves downward in a continual attempt to restore equilibrium by returning the sea's surface to its original flat condition. The attractive forces exerted by subsurface water on the water in the small ripples is sufficient to restore a rippleless condition.

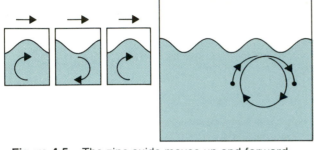

Figure 4.5 The zinc oxide moves up and forward as a crest passes and down and backward as a trough passes (a). The circle does not close, resulting in a forward motion of the zinc oxide (b).

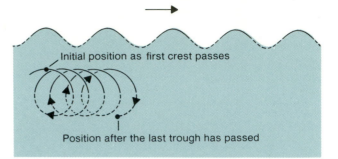

Figure 4.6 After a series of waves have passed, the zinc oxide has moved only slightly forward.

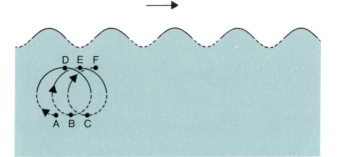

Figure 4.7 Points A, B, and C indicate the position of the water as the troughs pass; D, E, and F represent the position when the crests pass. The circles do not close; thus, the water moves slightly forward. This motion is termed mass transport.

Water Movement in Waves

Suppose that a marker such as zinc oxide had been added to various depths in a wave channel and the path of the gunk had been traced as a series of waves passed down the tank. Examination of such a tracing would show that the marker, and therefore the water slightly beneath the surface, moves upward as the crest of a wave passes and downward as the trough passes (fig. 4.5). In other words the water particles move up and slightly forward in the crest of a wave and down and slightly backward in the trough (fig. 4.6).

A closer examination of the tracing would reveal that the water describes a circular motion. The circle is not perfect, however, since it does not close completely. Rather there is a small forward motion of the water particles (fig. 4.7). This motion, though slight in relation to the number of waves that pass down the wave channel or over the surface of the sea, does move the water slightly forward with each passing wave. As a result the water exhibits an **orbital motion** as waves pass: it moves up and forward in each crest and down and backward in each trough with a very slight forward motion, since the orbits are not complete. The slight forward motion of the water is often called **mass transport.**

It is important to note that many waves must pass over a given point to produce only a very small forward motion of the actual water. Thus, it is the wave form, rather than water, that one observes moving through the sea or the wave channel. This motion is similar to that achieved by tying the end of a rope

to a door, holding the other end, and creating a series of waves by flicking the rope. The waves would be observed to move along the rope, but since one end of the rope is held, it is obvious that the particles that compose the rope are not moving forward; only the **wave form** is in motion.

If several drops of gunk were placed in the wave channel at various depths from the surface to the bottom, it would be found that the zinc oxide at a somewhat deeper level would make a smaller orbit than that nearer the surface. Depending upon the size of the waves generated, this orbit may or may not close completely. At the next depth, the zinc oxide would describe a flattened, closed orbit. The gunk at the deepest level would move forward as the crest passed and backward as the trough passed (fig. 4.8). The size of the orbit at any given depth is dependent upon the size of the waves; larger waves cause larger orbits to form with less closure and more forward

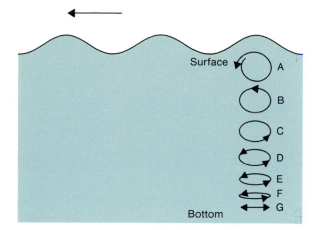

Figure 4.8 The movement of zinc oxide particles placed at different depths in a wave tank. Note that the shallowest particles describe an open orbital and are subject to mass transport. As the depth increases (C) the orbitals close completely; in deeper water (D-F) they flatten. The deepest particles merely move forward as a crest passes and backward as a trough passes. Only the particles at depths A and B exhibit mass transport.

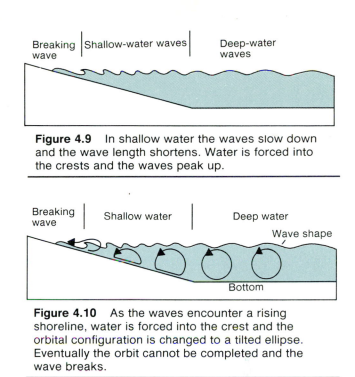

Figure 4.9 In shallow water the waves slow down and the wave length shortens. Water is forced into the crests and the waves peak up.

Figure 4.10 As the waves encounter a rising shoreline, water is forced into the crest and the orbital configuration is changed to a tilted ellipse. Eventually the orbit cannot be completed and the wave breaks.

motion than do smaller waves. Larger waves also cause the deeper water to describe circular open orbits rather than flattened closed orbits. In other words if larger waves were to be generated in the wave channel, the deeper water that had moved in flattened, closed orbits would begin to move in circular open orbits as the larger crests and troughs passed over.

If the wave channel is modified by creating a rising shoreline, the behavior of the water particles in shallow water can be investigated. When waves enter the shallow area of the wave tank, they begin to interact with the rising bottom, which creates "drag" on the waves. This friction causes the leading waves to slow down, while the waves in the deeper water maintain their original speed. Thus, the wave length is shortened. Simultaneously the rising bottom forces water to move from the trough to the crest, and the wave begins to steepen and "peak up" (fig. 4.9).

As water is forced into the crest, the orbital motion of the water particles changes and is forced into a tilted elliptical configuration (fig. 4.10). As the water becomes shallower, there is even less water available in the trough, the wave crest cannot be filled, and the symmetrical wave form cannot be completed. At this point the 1:7 ratio of wave height to length is exceeded. The wave becomes unstable, since the top of the crest has no support and the orbital motion of the water particles cannot be completed. The wave collapses and the water in the uncompleted orbits moves forward with the same speed as the unstable wave form, causing the wave to break, form surf, and move up the beach face. The area of breaking waves is called the **surf zone.**

Studies of waves in a laboratory wave channel show how individual waves, wave groups, and wave trains form, move, and ultimately break. These same processes occur in the sea, regardless of whether the waves were formed by wind, earthquakes, or the gravitational attraction of the sun and the moon.

Breaking waves

The surf zone

The swash zone

Wind Waves

As wind begins to blow across a flat sea, the energy from the wind is transferred to the water and waves are formed. This is due to the action of the wind, which creates friction as it moves across the water. This friction pushes against the water and if the wind is very gentle, a ripple will form; if the wind is stronger, larger waves will form from ripples.

Although only ripples may be formed initially, once the sea's surface is raised even slightly, the transfer of energy from the wind to the water is able to proceed more efficiently, since there is a steep side against which the wind can push. If the wind continues to blow, additional energy will be transferred and the ripples will get higher; that is, their wave height will increase. Eventually the wave's height-to-length ratio of 1:7 will be exceeded, and the wave will become unstable and break. When this occurs, the energy originally obtained from the wind will be transmitted into the calm water in front of the breaking wave.

This disturbance will raise a longer but initially lower wave, termed wind chop. Since wind chop is a longer wave, it can grow higher before it too becomes unstable. Should the wind continue to blow, this wave will break and since it is larger it will transfer a greater amount of energy to the calm water into which it breaks. This energy transfer will form lower but longer waves, referred to as fully developed seas. If the wind still continues to blow, the process will be repeated, and a very long, initially very low wave called swell will result.

As swell is formed and travels from the generating area, the waves are sorted. Waves with similar speeds and traveling in the same direction will move together as a rather uniform wave train. Since swell is long, it is a very stable wave and can travel for very long distances as a gravity wave, with minimal loss of energy. The type of waves that are actually formed depend upon the wind speed and duration, the distance over which the wind blows—termed the fetch—the distance from the generating area, and the bottom contours and depth of water.

Waves breaking on an offshore bar

Eventually swell will enter the relatively shallow waters of the continental shelf. When this occurs, the irregular swell of the deep ocean, formed by storms in countless areas of the sea, becomes organized as it passes over the shallow bottom and forms into a relatively uniform series of waves that move shoreward at similar speeds.

As the swell continues to move shoreward, the bottom will slow the incoming waves, the wave length will decrease, and water will be forced from the trough to the crest. As the wave destabilizes, the orbital motion of the water is interrupted, and the **tilted elliptical configurations** indicative of unstable waves are formed. Eventually the top of the crest cannot be supported by the trough, and the water, along with the wave form, moves in uncompleted orbits. This breaking wave forms surf.

If the waves break over an offshore bar, the water will generally deepen again and a new wave will form. This wave will then proceed shoreward, become unstable once again, and break on the beach face with a final loss of energy. The water will then travel up the beach in the **swash zone.** The net result of this process is the sudden release of the wind's energy on the beach face.

The energy that is released was accumulated from the wind during many storms and was stored and transported by the waves for hundreds, if not thousands, of miles. It becomes clear that the process of

The beach at Waikiki, Hawaii is famous for its spilling waves which are ideal for surfing.

sea building is relatively slow and occurs over countless miles of ocean, whereas the loss of this energy is rapid and is focused on a relatively small portion of the shoreline. Since the energy is released so rapidly in the surf zone, it is significantly higher than the energy in any one of the storms that shaped the surf into its final wave form. Calculations show that each time the height of a wave doubles, its energy increases by a factor of 4; thus, a tremendous amount of energy is stored in each wave. When the wave breaks, this stored energy is released rapidly. Breaking waves are therefore capable of moving tremendous quantities of sediment about in the surf zone (chapter 12).

The sand moved about by the surf is carried both along the shoreline and offshore to be deposited on the bottom. As each wave strikes the beach, it carries and deposits sand offshore, changing the configuration of the bottom. These changes, in turn, alter the characteristics of approaching waves and cause them to behave differently from their predecessors. With the waves continually changing the bottom and the bottom continually changing the waves, no two waves that strike a beach are exactly the same.

The bottom configuration forms either spilling or plunging breakers. Plunging breakers are formed as waves pass over a steep bottom that rises abruptly. These waves move in deep water and maintain their stability until they encounter the steep rise. At that point they will peak rapidly, become unstable, and plunge forward as breakers. Plunging breakers retain most of their energy up to the moment of breaking, when it is rapidly released.

Spilling breakers are formed as waves pass over a gently sloping bottom. These breakers build and destabilize slowly. When a wave does break, it tumbles down the gently sloping forward portion, losing its energy gradually. Waikiki Beach in Hawaii is well known for its spilling breakers. The bottom in this area is composed of a gently sloping coral reef that extends for over a mile offshore.

Figure 4.11 A series of sine waves, superimposed one atop the other, best illustrates the sea's surface.

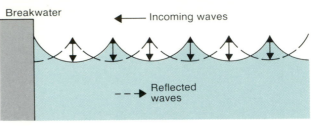

Figure 4.12 Waves reflected from a stationary object move back offshore and form a series of standing waves. Verticle arrows indicate the direction of the water movement.

The Waves of the Sea's Surface

Although waves in any given wave train are uniform as they leave a particular generating area, the surface of the sea is anything but orderly, since waves are generated in a great number of areas by variable winds. These waves, traveling from different generating areas, will meet at different angles, head-on, and so forth. In addition the wind seldom blows in a constant direction at a constant velocity. Hence, waves generated in a given area are themselves very variable and consist of waves of many different sizes, speeds, and shapes.

Sea waves, therefore, behave nothing like the orderly waves just discussed. Rather, each wave moves independently, with a size, shape, and direction different from its neighbors. The surface of the true sea appears to be, and actually is, best defined by superimposing a series of sine-wave trains one on top of the other (fig. 4.11). If this is done, there will be a point at which a number of wave crests meet and a very high wave will be formed. This wave will be transitory, however, since the individual waves that coalesced to form the "super" wave, often termed a rogue wave, will continue to travel in their own direction, at their own speed, etc., with the result that the "super" wave would dissipate. Similarly a series of troughs may coincide to form a "hole" in the sea's surface. This "super" trough would also be very transitory.

Rogue waves formed in this manner do indeed exist. Reliable observers on naval vessels, as well as on commercial ships, have reported huge waves, some over 30 meters (100 feet) high, that suddenly rear above neighboring waves. Since these rogue waves appear suddenly, dissipate rapidly, and presumably cause many seafarers to contemplate the hereafter, none have been photographed. These rogue waves and "super" troughs are, undoubtedly, responsible for many of the ships that have been lost at sea without a trace.

Reflection, Diffraction, and Refraction

When waves encounter stationary objects such as steep beach faces, vertical cliffs, or breakwaters, the waves may be **reflected, diffracted,** or **refracted.** When a wave train is reflected, the moving energy form is forced back upon itself to encounter additional incoming crests and troughs (fig. 4.12). For example, when waves encounter a breakwater, they are reflected back to set up a series of stationary standing waves in which the orbits of the standing waves encounter and modify those of the incoming waves. There is a vertical wave motion against the breakwater and a horizontal wave motion one-quarter of a wave length from the breakwater (fig. 4.13). Hence the wave energy of the incoming waves is reduced by the waves moving in the opposite direction.

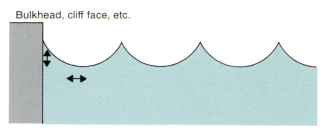

Bulkhead, cliff face, etc.

Figure 4.13 Reflection will result in a verticle movement against a stationary object and a horizontal motion one-quarter of a wavelength seaward.

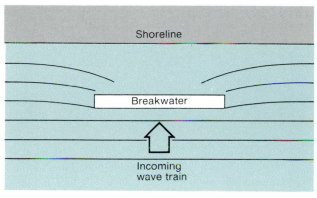

Shoreline

Breakwater

Incoming
wave train

Figure 4.14 Diffraction. The waves appear to bend around the breakwater.

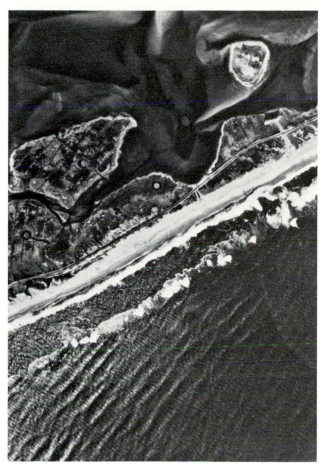

Refraction. Waves bend as they approach the shoreline.

When waves approach the end of the breakwater, they are also diffracted. Diffraction refers to the propagation of wave energy sideward as the wave extends into the area that is presumably protected by the breakwater (fig. 4.14).

Refraction is the bending of waves as they approach a shoreline. This occurs when waves move into shallow water and the leading waves interact with the bottom and slow down. Since some portions of the leading wave front travel in shallower water than other portions, there is a differential slowing of the wave front, and the crest bends. The differential speed and consequent refraction of the wave front cause the wave direction to change continually, which results in the wave front refracting until it becomes roughly parallel to the shoreline.

Shorelines are often irregular, with headlands projecting out into the sea. Since the subsurface contours often mimic those of the shoreline (fig. 4.15), refraction focuses waves toward the headland. This results in a large release of energy at the base of the headland and often causes significant erosion (chapters 5, 12).

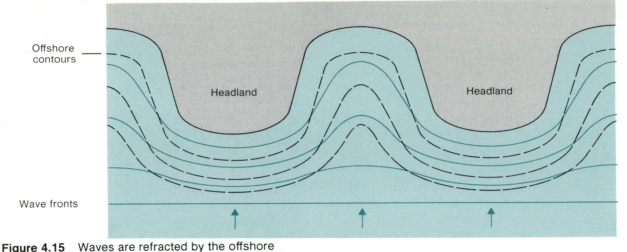

Figure 4.15   Waves are refracted by the offshore contours (a) and are focused on a headland (b).

The Tides

Tidal waves are nothing more than waves of extremely long wave lengths (table 4.1). Since they have long wave lengths, they can build to great heights without becoming unstable. In fact tidal waves are so long and so stable that they do not break. The crest of the tidal wave is known as high tide. As portions of the crest approach a landmass, the water levels increase and the tide is said to be rising. After the crest has passed, portions of the wave's following trough approach the land and the water level begins to fall.

The formation of these waves is governed primarily by Newton's laws of gravity. These laws show that although the gravitational attraction between two objects is directly proportional to the mass of the objects, the attraction decreases with distance and is actually inversely proportional to the square of the distance that separates the objects. This explains why the moon, though much less massive than the sun, exerts a greater gravitational attraction. Since the moon is only 390,000 kilometers (240,000 miles) from the earth, whereas the sun is 150 million kilometers (93 million miles) away, the moon exerts a far greater gravitational attraction, even though it is less massive than the sun.

The gravitational attraction of the moon and sun is counterbalanced by centrifugal forces. These forces arise as a result of the moon and the earth rotating about a common point, known to be 4,700

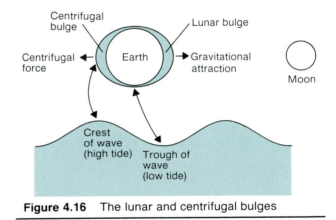

Figure 4.16   The lunar and centrifugal bulges

kilometers (2,900 miles) from the center of the earth. Since the earth and the moon rotate about a common point, the centrifugal and the gravitational forces are in balance at that point but are not equal on the surface of the earth. This causes the water to form a bulge, called a **lunar bulge,** on the side of the earth facing the moon.

The lunar bulge is due to the excessive gravitational attraction on that side of the earth. On the opposite side, there is an excessive centrifugal force as a result of which the water forms a **centrifugal bulge** on the side of the earth that faces away from the moon (fig. 4.16). These bulges are merely the crests of the tidal wave. Tidal bulges of this type would indeed be formed if the earth were perfectly smooth and covered by a uniform film of water. The presence of the continents, discussed in a later section, profoundly alters these tidal bulges.

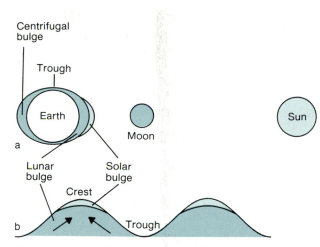

Figure 4.17 In the in-phase position the earth, moon, and sun are aligned and the solar bulge is superimposed upon the lunar bulge (a). Water, in this case, moves from the trough into the bulge (b).

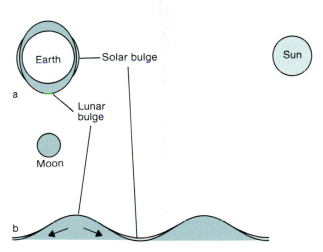

Figure 4.18 In the out-of-phase position the earth, moon, and sun are at right angles. Water is pulled from the lunar bulge into the trough, which is beneath the sun (a). This solar bulge decreases the height of the lunar bulge and increases the height of the trough (b).

The moon rotates about the earth in the same direction as the earth's rotation, with a complete lunar orbit taking 28 days; therefore, a given point on the earth must travel slightly farther than one rotation before it is again beneath the moon. This results in the tidal day's being 24 hours and 50 minutes long.

The tidal bulge is known to travel ahead of the moon rather than directly beneath it. This is due to two opposing forces: the friction of the earth and the

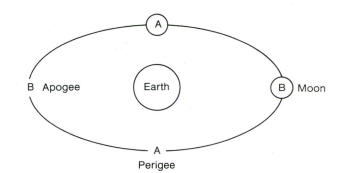

Figure 4.19 Since the moon's orbit about the earth is elliptical, it will be both closest and farthest from the earth twice each month. At point A, its closest point, the moon is in the perigee position; at point B, its farthest point, the moon is in the apogee position.

gravitational attraction of the moon. The friction of the earth tends to pull the bulge along at the speed of the rotating earth, while gravitational attraction tends to hold the tidal bulge beneath the moon. Hence the actual position of the tidal bulge is the net result of these two opposing forces. The continental land masses also impede and interfere with the movement of the tidal bulge.

There are several possible positions that the sun, moon, and earth may take in relation to each other. The most important are the in-phase and the out-of-phase positions. When the earth, moon, and sun are aligned, they are said to be in the in-phase position (fig. 4.17) at which point the **solar bulge** is superimposed on the lunar bulge. Water moves from the trough to the crests and there are higher high tides. Since there is less water in the troughs during these periods, the low tides are lower. These tides occur twice a month and are termed **spring tides.**

Twice a month the earth, moon, and sun are out-of-phase and at right angles to each other (fig. 4.18). During these periods water is pulled from the lunar bulge into the trough that is beneath the sun. As a result there is less water in the tidal crest and more water in the trough of the wave; therefore, there are lower high tides and higher lows during the out-of-phase position. These are called **neap tides.**

Other factors also affect the height of the tides. The moon travels about the earth in an **elliptical orbit** rather than in a circular one (fig. 4.19). Thus, the moon at two points is at its closest to the earth; this is called the perigee position. When the moon is in this position, it is 25,000 kilometers (16,000 miles) closer to the earth than when it is in apogee, which is the farthest point in its orbit. Since the gravitational attraction increases with decreasing distance, the tides at perigee are 20 percent higher, whereas the tides at apogee are 20 percent lower than normal. The moon is at apogee and perigee twice a month.

Twice a year the earth, moon, and sun are in-phase and the moon is in its perigee position. During these times the perigee tide reinforces the superimposed solar and lunar bulges, and large amounts of water move from the trough to the crest. As a result high tides are very high and low tides are extremely low. These tides are called **perigee in-phase tides.**

Also twice a year the earth, moon, and sun are out-of-phase and the moon is in apogee. During these periods the attraction of the moon is at its lowest, and the sun attracts water from the lunar bulge into the tidal trough. As a result there is the least amount of water in the crest, which causes very low high tides. Since there is a large amount of water in the trough during these periods, the low tides are higher than normal. At these times there is little difference in water height between high and low tide. The tides of these periods are called **apogee out-of-phase tides.**

Three additional types of tides—**diurnal, semidiurnal,** and **mixed tides**—may also be formed. Diurnal tides consist of one period of high water and one period of low water per day. Although uncommon, these tides are found in the Gulf of Mexico, Manila, and Vietnam. They are due primarily to the configuration of the ocean basins and the adjacent landmasses.

Semidiurnal tides have two equal or nearly equal high and low tides per day. The Atlantic Ocean has a semidiurnal tidal period. Mixed tides occur when each high and each low tide varies appreciably in height. In other words one high tide is significantly higher than the other. These tides occur in the eastern Pacific Ocean. Both mixed and semidiurnal tides are influenced by the angle or declination of the moon relative to the earth. When the moon is directly over the equator and the declination is least, mixed tides

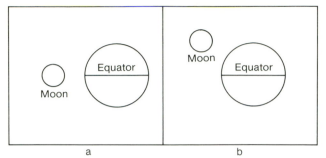

Figure 4.20 When the moon is directly over the equator and the declination is least (a), the height of the tidal bulges are fairly equal. When, however, the declination increases (b), mixed tides are common.

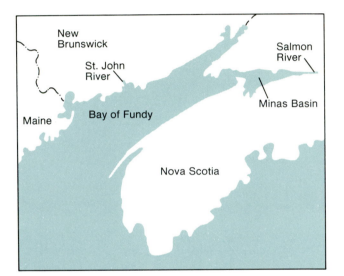

Figure 4.21 The funnel-shaped Bay of Fundy modifies and increases the height of the incoming waters on a rising tide. When this water encounters the land mass at the northeastern end of the bay, it is reflected back, further increasing the water level. Tidal bores are common in the Salmon River.

are more nearly equal; they are similar to semidiurnal tides. However, when the moon moves to the north or south of the equator, the declination increases and mixed tides result (fig. 4.20).

Coastal tides are actually oscillations that are caused by the tides moving through the deep ocean offshore the continental shelf. As the tidal wave approaches the shelf, it moves the water already over the continental shelf onshore. As this water moves over the rising shelf, the shelf acts as a wedge and

The tidal bore in the Salmon River, Truro, Nova Scotia

increases the wave height. Thus, the wider the continental shelf, the greater the coastal **tidal range.** For example, the shelf offshore Cape Hatteras, North Carolina, is relatively narrow and the tidal range in this area is 2 meters (7 feet) as compared with the tidal range along the coast of Maine, which is 3.5 meters (11 feet). Offshore Maine the shelf is considerably wider. As the landward portion of the coastal tidal wave approaches the coastline, it is reflected back. Then as it travels backward, it encounters additional incoming portions of the wave, and the height is further increased.

The shape of coastlines and embayments dramatically influences the height and characteristics of tidal waves. The classic example is the Bay of Fundy. This funnel-shaped bay (fig. 4.21) has a wide southern opening to the Atlantic Ocean—with no outlet on the north end—which is split into two narrow bays. As the wave enters the Bay of Fundy and travels northward, the coastline tends to constrict the wave front, which serves to increase its height. When the forward part of the wave encounters the northern end, it is reflected back to meet the water that is traveling into the bay, which further reinforces the wave height. As a result there is as much as a 10-meter (33-foot) difference between high and low tide in this bay.

On a smaller scale in other areas where rivers enter the ocean by means of long, funnel-shaped bays, the front of the incoming tidal wave will be restricted by the narrow channel and the shallow bay and river bottom. The restriction causes the wave front to increase in height and move upstream as a discrete wave called a tidal bore.

As the tides rise and fall, they form horizontal water movements known as tidal currents. In the deep ocean these currents are termed **rotary currents,** since they constantly change direction and make a complete cycle once during each tidal period. In coastal areas, however, the shoreline obstructs these currents and prevents their rotary movements. The result is that these coastal tidal currents move in one direction during parts of the tidal cycle and then reverse their flow. For this reason they

Box 4.1 The Tides and Animal Distributions

In coastal areas, tidal fluctuations directly influence the vertical distribution of organisms. This is particularly evident on rocky shorelines. In these areas there is a definite vertical zonation of organisms, with the common periwinkle snail (*Littorina*) occupying the uppermost portion of the rocks that are covered only at high tide. Below the *Littorina* zone there is invariably a zone dominated by the rock barnacle (*Balanus*) and beneath this is the zone occupied by the mussel (*Mytilus*).

Examination of the sedentary barnacles and mussels shows that those organisms that are higher up on the rock are generally smaller. These animals obtain food by filtering microscopic organisms from seawater. Since they are exposed to the air for longer periods, their feeding time is reduced. They therefore have a slower growth rate than their deeper, less exposed counterparts, which are able to feed for a greater portion of the tidal day. The periwinkle, on the other hand, is mobile, is able to regulate its position in relation to the different tidal cycles, and is therefore able to feed at a variety of different locations.

On sandy bottoms in the West Indies, the clam (*Donax denticulatus*) actually migrates up and down the beach face with the changing tides. This burrowing organism is able to live only in the swash zone, the portion of the beach consisting of saturated sand that is free of wave action. When the tide begins to rise, the clam emerges from its burrow and is carried shoreward by the incoming wave. As the water from the breaking wave travels up the beach, the water becomes calmer and the clam rapidly digs a new burrow in the saturated sediment above the surf zone. As the tide falls, this organism again emerges from its burrow and is carried farther down the beach to moister sand.

---

are called **reversing tidal currents.** They are called flood currents when they flow shoreward during a rising tide and ebb currents when they move seaward during a falling tide.

Reversing tidal currents have profound effects on the movement of river water. As the tide rises, the seawater moves upstream, encounters the lighter river water, and sinks beneath it. During this time the dense seawater flows upstream as a bottom-water current, while the less dense river water flows seaward as a surface current. As the tide continues to rise, surface flow will decrease until it equals the velocity of the subsurface flow.

At this point a convergence forms and marks the point where the flows are equal. A result of the equal but opposite water movements is that virtually no river water flows past the convergence. The rising tide will ultimately cause the surface flow to reverse, and the entire water mass will move upstream and accumulate. During this period no river water is discharged to the sea.

Immediately prior to high tide, the flood current will diminish and become weaker than the river current. Although subsurface water is still flowing upstream at this time, the convergence is moved seaward by the river water. As the tide begins to fall, the opposing upstream flow disappears completely, and the seaward flow increases as the accumulated water is discharged. When this water enters the adjacent coastal ocean, the river flow is dissipated and the river water is moved away from the mouth by coastal currents.

Tsunamis

The problems involved in finding a truly descriptive name for tsunamis has already been discussed (page 54). Since these waves are formed by disturbances within the earth's crust, they are more properly named seismic sea waves. The waves are formed by one of two processes: by the movement of the oceanic crust along a fault line (chapter 2) or by the eruption of a submarine volcano.

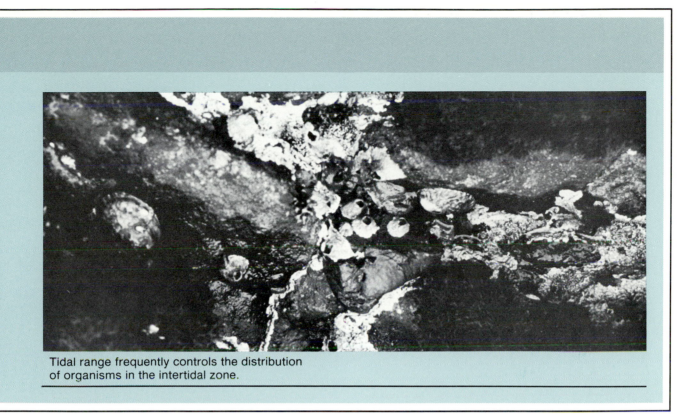

Tidal range frequently controls the distribution of organisms in the intertidal zone.

Because of movements of the earth's crust, tensions are created and build up along a fault. Eventually these tensions are reduced by an abrupt movement of the rock. When this occurs a huge mass of rock may suddenly rise or fall (fig. 4.22). Should this occur beneath the surface of the sea, the massive movement of the sea floor will cause a large displacement of the water in the overlying water column. This displacement causes the surface water to oscillate vertically, generating a series of very long waves.

Earthquakes commonly cause landslides. If a landslide occurs in a coastal area, large quantities of rock may suddenly fall into the sea, displace water, and form very long waves. Earthquakes frequently also trigger seaslides, submarine versions of landslides, in which huge quantities of neritic or pelagic sediment (chapter 3) are displaced. This sediment, in turn, displaces water and generates very long waves. Land and seaslides generally produce local tsunamis.

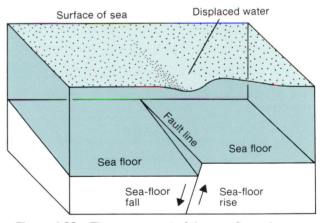
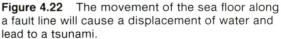

Figure 4.22 The movement of the sea floor along a fault line will cause a displacement of water and lead to a tsunami.

The aftermath of a tsunami

The eruptions of submarine volcanoes generate very destructive tsunamis as a result of their violent explosions, which displace huge amounts of water. In one instance the eruption of a volcano in the Pacific Ocean in 1883 produced tsunamis that killed more than 35,000 people.

Tsunamis, as they move away from the generating area, may have wave lengths of 241 kilometers (150 miles) and move at speeds in excess of 724 kilometers (450 miles) per hour. Initially they are very low waves, with heights that rarely exceed .3 to .6 meters (1 to 2 feet). Since they are so long, however, they have the potential for attaining great heights.

As these waves travel through the deep ocean, they are imperceptible, and passing ships are not only unaffected but are unaware of their presence. As they approach the continental shelf, however, the subsurface topography and rising shoreline cause the waves to slow down, the wave lengths shorten, and they peak up. When this occurs they are transformed into huge, unstable waves.

Just prior to the approach of the tsunami, hundreds of yards of sea floor are exposed as the coastal water is pulled into the crest. This adds to the wave height and further destabilizes the wave. At that point the entire wave mass begins to plunge forward as the orbital motion is disrupted.

Seiches

When long waves enter a protected area, they move rhythmically back and forth as they reflect off opposite ends of the land mass that surrounds the water body.

These oscillations, called **seiches,** are essentially standing waves that are similar to those formed when waves reflect from a breakwater. With seiches, the pattern of standing waves is composed of nodes, where the height of the wave remains constant, and loops, where the water moves vertically (fig. 4.23). The nodes, loops, and surface water maintain a set position, but strong subsurface currents are generated in order to support the shifting wave form. Long waves other than tsunamis will also cause seiching in protected bays of the proper configuration.

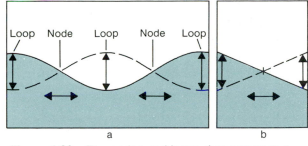

Figure 4.23 The nodes and loops that compose a seiche

Seiches may also be formed by a change in atmospheric pressure. For example, when a low-pressure air mass (chapter 10) moves into an area, the water may rise and begin to move back and forth in the basin in response to the reduction in air pressure. This is common on inland water bodies such as Lake Erie.

Summary

All waves, from the smallest ripple to the most destructive tsunami, have common characteristics. They all have crests, troughs, wave heights, lengths, and periods. In addition the water particles that compose the waves all move in identical orbital patterns: up and forward in the crest and down and back in the trough. It is only when the wave becomes unstable that the orbital motion is interrupted and the water particles begin to move at the same speed as the moving wave form.

Breaking waves release a tremendous amount of stored energy on a beach face. This energy moves the sand about and changes the configuration of the bottom. As the bottom configuration is changed by the waves, it changes the characteristics of incoming waves. This interaction between the waves and the bottom results in the beach face having an ever-changing wave pattern.

The longest waves in the sea are the tides, which are formed as a result of the gravitational attraction of the moon and the sun on the water of the earth. Even though the moon is much smaller than the sun, since it is considerably closer to the earth, it has the major effect on tides. The positions of the moon and the sun in relation to the earth are responsible for the spring and neap tides, as well as for the apogee and perigee tides. The configuration of the ocean basins and the declination of the moon are the major factors involved in the formation of diurnal, semi-diurnal, and mixed tides.

The magnitude of the tides is profoundly influenced by the configuration of the coastline. This is particularly evident in the Bay of Fundy, where there is as much as a 10-meter (33-foot) difference between high and low tide.

Tsunamis, more appropriately called seismic sea waves, are generated by movements of the earth's crust or by the eruption of subsurface volcanoes. These phenomena cause massive disturbances and displacement of the water to form very long waves. Initially these waves are low, but since they are long, they are capable of building to tremendous heights. This occurs as they travel onto the continental shelf. As the waves build they become unstable and break on a coastline, causing massive destruction.

Review

1. Diagram the parts of a typical wave train.
2. Distinguish between tidal waves and tsunamis.
3. How are lunar and centrifugal bulges formed?
4. Explain how swell is formed by wind blowing over a flat, calm sea.
5. What happens to the subsurface water as a series of waves pass over a given area?
6. Explain why a wave breaks.
7. Diagram the position of the earth, moon, and sun during a spring and a neap tide.
8. How are apogee and perigee tides formed?
9. What is a seiche?
10. Explain diffraction, reflection, and refraction.

References

Kinsman, B. (1965). *Wind Waves: Their Generation and Propagation on the Ocean's Surface*. Englewood Cliffs, N.J.: Prentice-Hall.

Phillips, O. M. (1966). *The Dynamics of the Upper Ocean*. England: Cambridge U. Press.

Russell, R. C., & MacMillan, D. M. (1954). *Waves and Tides*. England: Hutchinson.

For Further Reading

Bascom, W. (1980). *Waves and Beaches*. New York: Doubleday.

Clancy, E. P. (1968). *The Tides: Pulse of the Earth*. New York: Doubleday.

Redfield, A. C. (1980). *The Tides of the Waters of New England and New York*. Mass.: Woods Hole Oceanographic Institute.

5

Beaches, Coastlines, and Coastal Sediment

Key Terms

beach
beach face
coastline
depositional shoreline
erosional shoreline
glaciated coastline
high-energy environment
longshore current

low-energy environment
moderate-energy environment
primary coastline
secondary coastline
slope
stable shoreline
unglaciated coastline

The major and most obvious effects of waves and tides are observed along the coastal strip in the zone between the high- and low-tide levels. It is in this area that the waves, formed far offshore, rapidly expend their stored energy as they break upon the shoreline. This energy is used to carry sediment from the beach to the surf zone and from the surf zone to the beach. Much of the sediment originates as a result of erosion of the adjacent land masses by wave action or is carried to the coastal sea by river flow or surface runoff during rain storms and other such occurrences.

Often well-defined currents are formed in the surf zone. These currents are capable of transporting considerable amounts of sediment for long distances along a coastline. As this sediment is carried along, it is continually sorted and mixed with other sediments that are also brought into the coastal sea at different points along the coast by erosion and river flow. Ultimately, depending upon the local conditions, these sediments will either be deposited offshore as neritic or pelagic sediments or will settle out on a beach.

The residence time of sediments that have been deposited on a beach is often very short, since they may be resuspended almost immediately and carried elsewhere. The deposition and subsequent removal of these sediments are profoundly influenced by the ever-fluctuating tidal levels. As the tide rises, waves will strike and remove or deposit sediments higher on the beach. As it falls, the waves will deposit or remove sediments that are lower on the beach, with the result that the areas of deposition and removal are continually shifting in response to the changing levels of the tide.

In addition the actual level of the shoreline may be continually changing in response to tectonic and isostatic movements of the earth's crust or changes in the level of the sea. This too exposes different portions of the shore to wave attack, albeit on a much longer time scale.

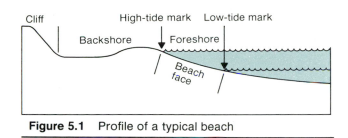

Figure 5.1 Profile of a typical beach

Beaches and Shorelines

Beaches are defined as areas that extend from the neap tide mark landward to the bases of cliffs or to zones of permanent vegetation. A beach may be divided into the foreshore, or shoreline, and the backshore (fig. 5.1).

The foreshore is that portion of a beach that is periodically covered and exposed during high and low tides. Thus, the foreshore is immediately landward of the surf zone. The section of the foreshore that is exposed to wave action slopes toward the sea and is termed the **beach face.**

The slope of the beach face is directly related to sediment size and, since the size of the sediment is controlled by wave action, the slope may be used as an indication of the wave energy that is released on any given shoreline. Generally, the greater the slope, the coarser the sediment and the greater the wave action. Consequently beaches composed of fine sand have slopes of less than five degrees; beaches consisting of pebbles have slopes of approximately fifteen degrees; and cobble beaches slope approximately twenty-five degrees.

Dunes are often found landward of the beach on
the east coast.

The backshore, the area that is commonly referred to as the "beach," extends landward from the
high-tide mark to the base of a cliff, dune, or zone
of permanent vegetation. This area receives water
only from splashing waves or during storms. The
backshore consists of one or more berms that are
either flat or slope gently landward. Where cliffs are
absent, dunes built by windblown sand are generally
found behind the berm (fig. 5.2).

Frequently, a submerged ridge of sand, known
as a bar, may run parallel to the shoreline offshore
the surf zone. Generally sand moves from the beach
face to the bar in the winter and from the bar to the
beach face in the summer. This movement of sand is

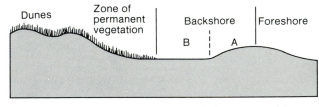

Figure 5.2   The berms of the backshore may either
slope landward (A) or be flat (B).

due to the difference between summer and winter
waves and influences the **slope** of the beach face, as
well as the size of the bar.

Strong winter winds generate large waves that
break upon the shoreline with great frequency and
force. As a result the beach face becomes saturated
with seawater. Additional incoming waves break and

Winter waves customarily steepen a beachface.

pick up sand, which they carry up the beach face toward the berm. The leading edge of this water, containing the suspended sand, reaches the landward edge of the beach face, stops, and deposits the sand on the crest of the berm. Since the beach face is saturated, the water must return to the surf zone over the surface. As it travels seaward, the water picks up sand from the beach face and carries it to the surf zone. It encounters incoming water carrying its load of suspended sand, loses velocity, and deposits the sand on the offshore bar. Thus, in the winter, the berm is built by the sediment carried ashore by the incoming water. When this water returns offshore, it picks up sand from the beach face causing it to steepen. This sand is then deposited on the bar which becomes higher.

In the summer the winds and waves are gentler and the beach face rarely becomes saturated. As these smaller summer waves travel onshore, they remove the finer sand from the bar and carry it to the beach face. When these waves break they travel ashore. The water sinks into the unsaturated beach face and returns offshore as subsurface flow. The sand carried from the bar is deposited upon and remains on the beach face.

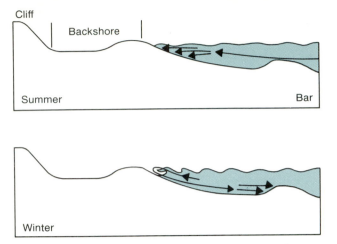

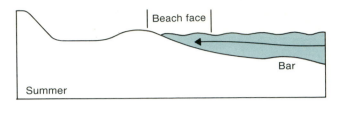

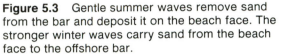

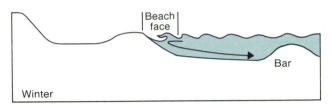

Figure 5.3 Gentle summer waves remove sand from the bar and deposit it on the beach face. The stronger winter waves carry sand from the beach face to the offshore bar.

Figure 5.4 Sand moves from the bar to the beach face in the summer and, as a result, the beach face is broad with a gentle slope and the bar is lower. In the winter the reverse occurs—sand moves from the beach face to the bar. The bar builds while the beach face has a steeper slope. Arrows indicate the direction of the sand movement.

As a result of these processes, the summer beach has a gentle slope and consists of fine sand that has been removed from the bar (fig. 5.3). In the winter the fine sand is removed from the beach face, and the winter beach has a greater slope and consists of coarser sediment (fig. 5.4). If, however, there is a paucity of fine sand in a given area, the dominant sediment will be coarse sand. In these cases the gentler summer waves will remove the available supply of fine sand from the beach face, while the larger winter waves will return coarser material. When this occurs, the summer beach may actually be steeper than the winter beach.

In addition to moving vertically between the bar and the beach face, sediment is also transported horizontally along the beach in the surf zone. The current that transports this sediment is known as the **longshore current.** This current is formed when the waves strike the beach face at an angle. If the waves are approaching the beach from the left, they will break at an angle, pick up sand, and move it directly offshore into the surf zone. The sand will then be

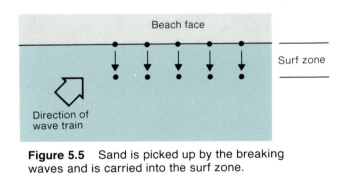

Figure 5.5 Sand is picked up by the breaking waves and is carried into the surf zone.

driven back to the beach face by another incoming wave (fig. 5.5). This sediment replaces sand that was removed from that portion of the beach and carried farther down the beach by a previous wave. The sediment deposited is soon removed by another wave, which resuspends it, carries it into the surf zone, and redeposits it slightly farther down the beach to replace another sand grain that was removed previously (fig. 5.6). Thus, the sediment moves down the beach in a zigzag motion (fig. 5.7). The size of the sediment that is carried by the longshore current is dependent upon the energy of the waves that strike the beach at that particular time.

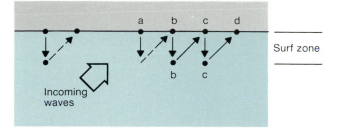

Figure 5.6 A sand grain, picked up and moved into the surf zone, is carried back to the beach by the next incoming wave. Since the wave is approaching the shoreline at an angle, the sand grain is carried back and strikes the beach to the right of its original position. The result is that sand grain a replaces b, which replaces c, etc.

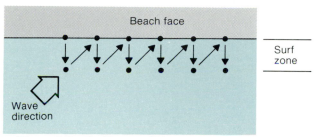

Figure 5.7 Sediment will be moved from the beach to the surf zone and back to the beach again. Since the waves transporting the sediment strike the beach face at an angle, the sediment is transported down the beach (in this case to the right) in a zigzag path.

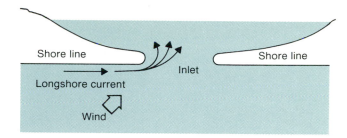

Figure 5.8 When the longshore current encounters and enters an inlet, it will disperse. As a result the velocity will decrease and sediment will be deposited.

There are areas along any beach where, for a variety of reasons, the waves strike most strongly. In these areas, the finer sands will be removed and the slope of the beach face will be steep and consist of coarser sand, pebbles, cobbles, or even bedrock. The finer sediments will, ultimately, be deposited in calmer, more protected areas, resulting in a beach face with a gentler slope.

Silts and clay-sized particles are deposited only in the calmest areas. These sediments often form extensive mud flats that eventually build high enough to be exposed at low tide. When this occurs, marsh grasses or mangroves begin to invade. These plants facilitate the deposition of sediment by slowing down the incoming water, causing the waterborne sediments to be deposited at a more rapid rate. In these instances the entire area may be rapidly converted into an extensive salt marsh or mangrove forest. This phenomenon is quite common along the mid-Atlantic and Gulf coasts, which are often indented and/or protected by barrier beaches.

As a result of the movement of sediment along a beach, shorelines are generally considered to be **erosional, depositional,** or **stable.** In order to determine the status of a particular shoreline, observations must be made over a long period of time, since the deposition and removal of sediment is often seasonal. Thus, someone observing a coastline only in the winter, when sediment accumulates on the offshore bar, may erroneously conclude that the shoreline is eroding when over the long term, it may, in fact, be accumulating sediment and be building.

An erosional shoreline, as the name implies, is one where there is a net removal of sediment. For example, should an offshore subsurface bar be low or absent in a particular area, the waves, rather than first breaking and expending their energy on the bar (chapter 4), would break directly on the beach face. As a result there would be a net removal of sediment from the beach face, which would steepen and cut back. Should this occur for a long period on a barrier beach, or should the area be struck by a hurricane or severe storm, a new inlet could form.

Depositional shorelines result when more sediment is deposited than is removed. This frequently occurs when the longshore current encounters and disperses into an inlet. As it disperses, the current loses velocity and the sediments that it had been carrying are deposited. This causes the inlet to shoal and narrow (fig. 5.8). On the barrier beach on the south

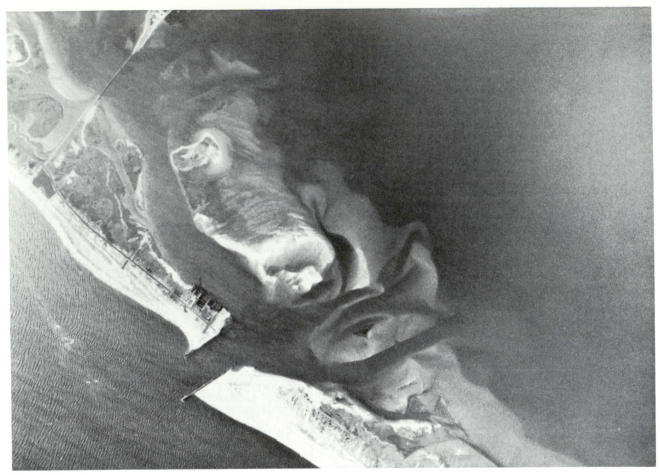

A shoaling inlet. The sediment has accumulated to such an extent that it is now permanently above sea level and has formed a tidal delta.

shore of Long Island, New York, a lighthouse was constructed at the mouth of an inlet in the late 1800s. As a result of the long-term deposition of sediment, this lighthouse is now more than two miles from the present inlet.

Stable shorelines are rare and occur only when the removal of sediment by the waves and currents is exactly balanced by deposition. In reality probably no shoreline is ever truly stable over the long term.

Shorelines and beaches are only parts of **coastlines,** or the coastal zone, which encompasses the area from the neap-tide mark to the landward limit that is inundated by storm waves. Coastlines may be rocky and mountainous, as along the Pacific and the north and northeastern Atlantic coasts, or gently sloping as along the mid-Atlantic coast.

Primary and Secondary Coastlines

Coastlines are also frequently classified as **primary** or **secondary.** Primary coastlines are new coasts that have only recently come into contact with the sea and become exposed to the action of the waves and tides. These coasts result from one of three processes: tectonic activity, isostatic adjustment, or the action and melting of glaciers.

As noted in chapter 2, when crustal plates diverge, new seas may form, while the convergence of plates may result in the formation of volcanoes and volcanic islands. Isostatic adjustments, on the other hand, may cause land masses to rise or subside. In addition during the Ice Ages, glaciers removed large quantities of sediments from the continents and carried this sediment with them as they moved southward. When these glaciers halted and began to melt

The sea stack at Tickle Bay, Newfoundland. Sea stacks are often formed as a headland erodes. Since sea stacks are composed of more durable materials they are resistant to erosion and their presence can be used to infer the past position of a headland.

and recede, the sediments, consisting of boulders, rocks, pebbles, sand, and silt were left behind as terminal moraines. Long Island, New York and Cape Cod, Massachusetts, are examples of such formations. As they melted, the glaciers released large quantities of water that entered the marine environment and caused sea levels to rise. The rising sea level inundated previously exposed lands, flooded river valleys, and so forth. All of these activities resulted in the formation of primary coastlines; although these coastlines have not yet been fully shaped by the sea, they are generally sites of active erosion.

Once such an area becomes exposed to the sea, refraction of the wave train will focus the waves' energy on headlands, if present, or if headlands are not

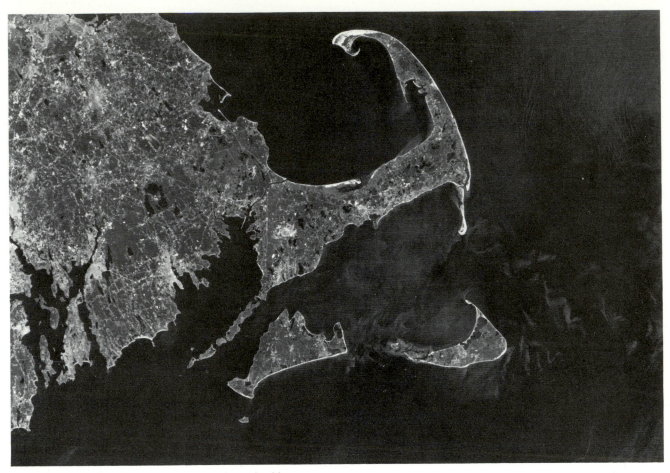

Cape Cod was formed by sediments deposited by a glacier.

present, along the entire expanse of shoreline. As waves attack a primary coastline, the finer sediments are carried into the surf zone, while the larger material is mechanically reduced. Once this material becomes small enough, it too will be moved about by the waves. The longshore current will then begin to move these materials along the shoreline. Eventually these sediments form secondary coasts when they are deposited as bars, barrier islands, and other such formations.

Secondary coastlines are defined as areas that have been shaped by marine processes. Thus, secondary coastlines are often formed from sediments that have been derived from primary coasts. Figure 5.9 illustrates the relationship between the two

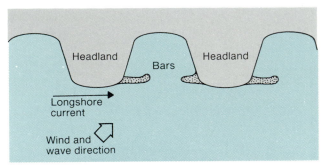

Figure 5.9 A consolidated headland is an example of a primary coastline. As the rocks are broken down into smaller sediments, they will be carried by the longshore current. When the current disperses into an inlet, embayment, etc., the sediment will be deposited. When sufficient sediment has been deposited, a bay mouth bar will be formed. Such a bar is an example of a secondary coastline.

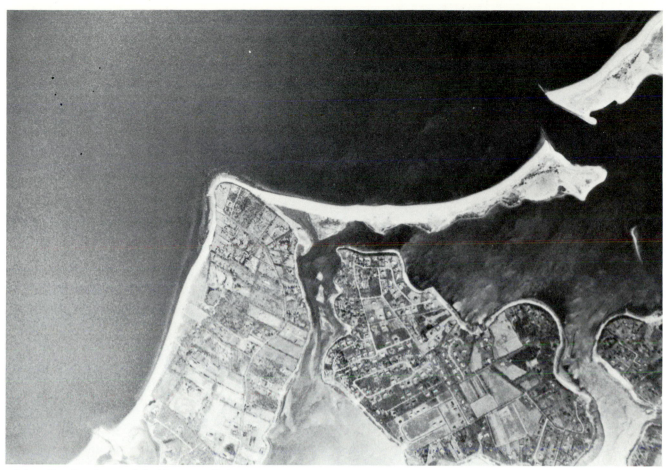

A headland and associated bay-mouth bar

Table 5.1 Types of Coastlines	Coast	Origin of Sediment	Source
	Primary	Terrigenous	Erosion
	Secondary	Terrigenous or Biogenic	
	Type A	Terrigenous	Erosion, Rivers, Runoff
	Type B	Biogenic	Marine Organisms

coastlines. For example, sediments eroded from a headland are carried into the surf zone by wave action and are moved along the beach by the longshore current. When this current encounters an indentation in the coastline, it will disperse, lose velocity, and deposit the sediments at the tip of the headland. Eventually these sediments become permanent above-water features known as bay-mouth bars. Secondary coastlines may also be formed by marine organisms. Coral reefs, therefore, would also be considered secondary coastlines. Secondary coastlines, then, may consist of either terrigenous or biogenous sediments (chapter 3). To distinguish between the two, they may be termed Type A and Type B secondary coastlines or merely Type A and Type B coasts, respectively. Table 5.1 summarizes the different coasts.

A rocky shoreline. A mat of marine vegetation is covering the rocks in the foregound.

Types of Coastal Sediment

The majority of sediments found along a primary and a Type A secondary coast are terrigenous in origin and consist of whatever materials compose the adjacent landmasses. Since the continents were formed as the less dense granites moved to the earth's surface (chapter 2), the sediments derived from primary coastlines are generally the by-products of mechanical weathering of parent granite—light-colored quartz and feldspar. The associated secondary coasts also consist of this type of sediment. The materials formed by the weathering of volcanic basalt, on the other hand, are dark and are rich in iron and magnesium. Thus, the beaches associated with volcanic islands and with continental coastal areas where basalt is exposed to wave attack are frequently a black-green in color.

The sediments that compose Type B secondary coasts often consist of the shells and shell fragments of organisms that live in the shallow coastal waters. Many beaches in Florida and Mississippi consist, primarily, of the shells of small clams, while the beaches on eastern Long Island, New York, are piled high with the shells of limpets after storms. On Long Island, shell beaches are ephemeral, since the shells are moved back offshore soon after the storm passes. In

The skeletal bases of coral

Fringing reef

tropical areas the beaches often consist of the protective covering of various types of *Foraminifera* (chapter 3). As a result, these beaches are generally pure white, or, as in the cases of the beaches in Bermuda, a light pink, since the coverings of the *Foraminifera* that compose those beaches are pinkish red.

Type B secondary coastlines may also be composed of coral. Three types of coral reef are found in the world's tropical and subtropical seas: fringing reefs, barrier reefs, and atolls. Figure 5.10 illustrates these reefs.

Fringing reefs are always found contiguous with a landmass. These reefs are formed in the shallow sea, where rocks provide a point of attachment for the coral. Since fringing reefs are, by definition, contiguous with a landmass, lagoons are not present. Fringing reefs are found in the coastal waters of the Yucatan Peninsula, Mexico.

Barrier reefs are always separated from the coastline by a lagoon. These lagoons generally range from 100 meters (328 feet) to 3 to 4 kilometers (2 to 2.5 miles) in width, although the Great Barrier Reef of Australia has a lagoon that ranges from 11 to 65 kilometers (7 to 40 miles) in width.

Atolls are incomplete rings of sandy islands that have formed on top of coral reefs. The lagoons of atolls are within the ring. Atolls are much more common in the Pacific than in the Atlantic Ocean.

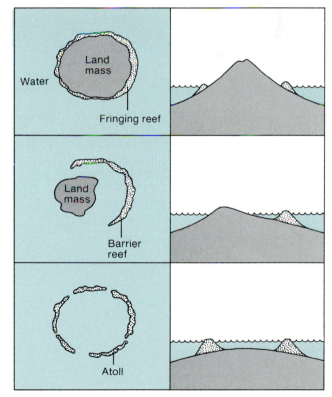

Figure 5.10 Types of coral reefs

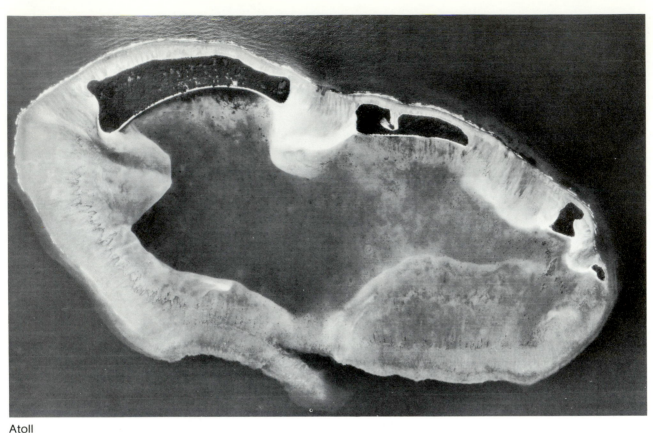

Atoll

A beach is composed of whatever material is available on or in the adjacent landmasses or in the coastal water. The size of the sediment that is found on any particular beach is dependent upon the energy expended by the breaking waves and the currents. Coasts are often considered to be **high-**, **moderate-**, and **low-energy environments** on the basis of the energy released by these breaking waves. A high-energy environment would have steeply sloped shorelines consisting of boulders, pebbles, etc. The energy released along a coastline is often dependent upon the fetch (chapter 4) and the topography of the surrounding landmasses, as well as on the subsurface topography.

The Origin of Coastal Sediment

The majority of the sediments found along coastlines are terrigenous in origin. They are either derived from the erosion of primary coastlines or have been carried into the marine environment by rivers or surface runoff during rain storms. As noted previously, the sediment that is eroded from primary coastlines is generally derived from the materials that compose the continental landmasses. Initially, upon entering the coastal sea, this material is poorly sorted. The finer materials are carried away almost immediately. The sand may be carried offshore to form the neretic sediments of the continental shelf, slope, and rise, while the clays and silts may be carried even farther offshore before they settle out to contribute to the pelagic sediments of the deep-ocean floor. On the other hand, these materials may be transported

Box 5.2 The Formation of Coral Reefs

Two major theories have been developed in order to explain the formation of the three types of coral reef. One, proposed by Charles Darwin, considers subsidence of the landmass to be the underlying factor; the other, formulated by Charles Daly, is based upon sea level changes during glacial and interglacial periods in the earth's history.

Darwin believed that both barrier reefs and atolls began as fringing reefs in the shallow coastal waters offshore volcanic islands. As these islands subsided (chapter 2), the coral continued to grow, one atop the other, and kept pace with the sinking land. As a result of this continual growth, the coral remained only a few meters beneath the sea's surface.

Since coral utilizes microscopic oceanic plants (chapters 8, 9) as a food supply, the coral on the seaward side of the reef was in a favored position, whereas the coral on the landward side of the reef received less food and eventually perished. When the coral on the landward side died, the reef became detached from the landmass and the lagoon was formed. Should the landmass continue to subside and eventually sink beneath the sea, the coral would continue to grow, resulting in an atoll.

Daly contended that in the periods between glaciers, fringing reefs were common in the warmer seas. During glacial periods, however, seas cooled and much of the coral was killed. Moreover, large quantities of water were tied up as glacial ice and so sea levels fell. It is estimated that during glacial periods sea levels fell by at least 60 meters (197 feet).

During these periods, the surf cut platforms in the exposed rock and in the dead coral that formed the exposed reefs. When the glaciers melted, sea levels rose as the water warmed. Coral once again flourished and, due to a ready food supply, the seaward edges of the platforms supported the most rapidly growing coral. The coral on the seaward side of the platform, nourished by its ample food supply, was able to keep pace with the rising sea level, whereas the coral on the landward side perished. Thus, barrier reefs and accompanying lagoons were formed. According to Daly, atolls would have formed on the submerged platforms that surrounded drowned islands.

In actuality, some reefs probably formed by subsidence, while others are formed in response to a rising sea level. Thus, both theories are right, depending upon the specific reef in question.

---

along the shoreline by the longshore current. When this occurs, the sand forms or augments secondary Type A coasts, such as barrier beaches. In this case, finer clays and silts would settle only in the calmest, most protected areas, where they form mud flats, contribute sediment to salt marshes, etc.

Surface runoff carries sediments from the immediate area into the sea, whereas large rivers can transport sediments from a wide geographic area into the coastal ocean. The sediments that are carried by surface runoff are generally the finer silts and clays that are able to remain in suspension for considerable periods of time. It is these materials that often cause bays to appear muddy after a heavy rainfall.

Ultimately these sediments are either carried offshore or along the shoreline and are distributed in a manner similar to the distribution of fine materials derived from primary coasts.

The transport of sediment to the coastal sea by rivers is often seasonal and, in the temperate zone, generally corresponds to the increased volume of river water generated by spring rains and the melting of snow and ice. During the majority of the year, sediments carried by rivers tend to become trapped in the river system. Although they do move downstream suspended in the river water or as the bed load, when the river approaches the sea, tidal influences become predominant and these materials are carried back upstream on the rising tide. Frequently the incoming tidal water also resuspends previously deposited material and carries this material upstream as well.

During periods when the river receives rain and melt water, its volume and speed is increased. At these times significant amounts of sediment do enter the coastal waters. Storm events also transport large quantities of sediment into the sea. For example, in 1973, as a result of tropical storm Agnes, the Susquehanna River supplied thirty times its normal annual sediment load to the Chesapeake Bay. In fact, it has been estimated that the combined effects of Agnes and an unnamed storm that occurred in 1936 have supplied approximately 50 percent of all of the terrigenous sediment that has entered the Chesapeake Bay system during this century.

When riverborne sediments do enter the sea, most of the materials remain in suspension and follow pathways offshore or along the coast similar to those described for sediments derived from primary coasts. In some cases, such as the Frazer River in British Columbia and the Mississippi River, more sediment enters the marine environment than can be dispersed. When this occurs, the excess sediment is deposited at the river's mouth to form a delta or bar.

Sediment Dispersal

The tendency of sediment to be removed from a given area is dependent on six factors: (1) the size of the sediment itself, (2) the amount of sediment that enters the sea, (3) the tidal range, (4) the strength of

The Chesapeake Bay. Two storms have supplied fifty percent of the sediment that has entered this bay since 1900. The Susquehanna River empties into the north end of the bay. During tropical storm Agnes the river discharged 30 times its annual sediment load to the bay.
The Delmarva Peninsula separates the Chesapeake Bay from the Delaware Bay to the east. Note the extensive barrier beaches extending from north to south offshore New Jersey, Delaware, Maryland and Virginia.

Box 5.3 Sediments and Coastal Productivity

Suspended sediment tends to physically block sunlight and prevent it from penetrating deeply into the water column (chapter 8). Moreover the presence of very fine sediments in the water can clog the gills of fish and cause a high mortality. Should these materials settle to the bottom, they may bury the eggs of marine organisms, young shellfish, etc. The excessive input of sediment can, therefore, decrease the productivity of coastal areas significantly.

Coastal areas are in great demand for homes, commercial establishments, and other structures. Often, during the construction of these structures, little care is taken to control erosion, and large amounts of sediment find their way into the coastal environment and adversely affect the marine life.

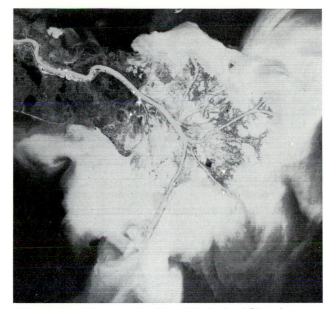

The delta at the mouth of the Mississippi River has formed because of the relatively weak currents and low tidal range in the Mississippi Sound.

the tidal currents, (5) the energy expended by the breaking waves, and (6) the strength of the coastal currents. Moreover the topography of the coastline will also influence the deposition or removal of sediments.

Since primary coastlines are sites of active erosion, it can be assumed that the tidal range, currents, and wave energy are sufficient to remove sediments. The rate at which the sediments will be carried away from the area is dependent on the sediment size and the coastal currents. Fine sand and smaller particles will, most likely, be carried away immediately, while the larger materials will be mechanically reduced prior to transport. The sediments carried by surface runoff are very fine and will also be readily transported from the site of input.

The majority of sediments transported by most rivers are generally trapped in the tidal portion of the river and are only discharged seasonally or during storms. The limited amount of sediments that are discharged are generally distributed along the shoreline by the longshore current or are carried offshore by coastal currents. When large quantities of sediment are discharged, they may accumulate at the mouth of the river as a delta. For example, the Mississippi River discharges approximately 300 million tons of sediment into the Mississippi Sound annually. This large input of sediment, in relation to the low tidal range and relatively low wave energy in this particular sound, has allowed the Mississippi River Delta to form.

On the other hand the Columbia River, which forms the boundary between Oregon and Washington, discharges approximately 20 million tons of

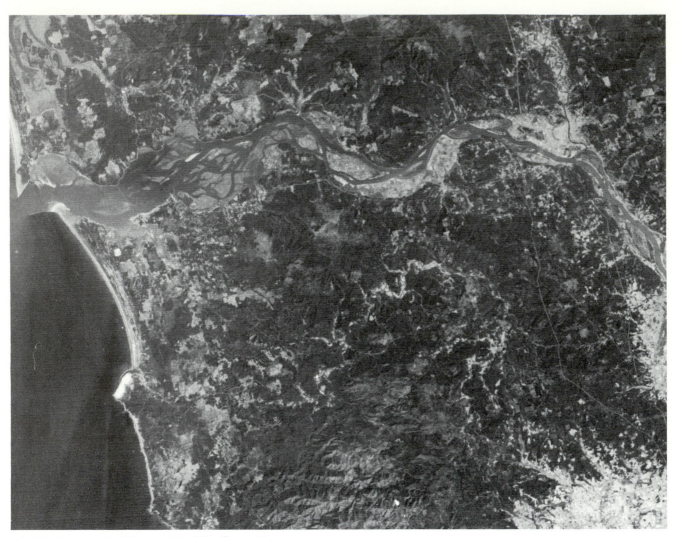

A delta is absent at the mouth of the Columbia River. This is due to the strong currents and large tidal range which distributes the sediment along the coastline to form rather extensive sandy beaches and sand dunes.

sediment per year to the Pacific Ocean; yet a delta is absent at this river's mouth. This is due to the large tidal range and strong wave action in this area. This sediment is removed from the immediate vicinity of the river mouth and is distributed along the coast to form the largest beaches and dune systems found along the entire Oregon–Washington coastline.

Slightly farther to the north, the Fraser River flows into the Puget Sound. A delta has formed at the mouth of the Fraser River, even though the Fraser discharges less sediment than the Columbia River. The presence of this delta may be attributed to Vancouver Island, which is located to the west. This island reduces the wave strength at the mouth of the Fraser River, thereby minimizing sediment transport.

A delta has formed at the mouth of the Fraser River
(upper right) in British Columbia due to the
protection provided by the adjacent landmasses.

The Major Coastlines of North America

The coastlines of North America may be divided geographically into the Atlantic, the Pacific, and the Gulf coasts. Since the continental plate (chapter 2) is moving westward, the Pacific coast is at the leading edge of this plate and is tectonically more active than the Atlantic coast. The Atlantic and Pacific coasts may be further subdivided on the basis of their exposure to glacial activity. During the ice ages, glaciers moved as far south as the states of Washington and New York. As they moved they scraped the surface sediment from the continents. When they halted and began to recede, this material was deposited as terminal moraines. As a result, Washington, New York, and the states and provinces to their north have only a thin veneer of sediment covering the bedrock. Along the coastal areas, the bedrock is generally at the surface and exposed to the sea. These are considered to be **glaciated coastlines.** Since the coastal sediments consist of durable bedrock, beaches are scarce in these areas, and, where present, often consist of pebbles or very coarse sand.

The coastlines to the south of New York and Washington have not been exposed to glacial activity and are, therefore, termed **unglaciated coastlines.** Bedrock is not exposed and the sediments are loosely packed or unconsolidated. As a result, these materials are moved about quite easily by the waves and currents. Unglaciated Atlantic coastlines are generally broad plains that slope gently to the sea, whereas those on the unglaciated Pacific coast are characterized by cliffs and coastal mountains.

The Pacific coast is characterized by high-energy conditions, since the breaking waves are large and release considerable energy as they break upon the shoreline. As a result considerable amounts of large-sized sediment are moved about, with much of this sediment being drained off into submarine canyons. Thus, broad beaches are less numerous on the unglaciated Pacific coast and are usually found near a plentiful source of sediment, such as down-current of large rivers. The Atlantic coast is characterized by moderate-energy conditions; there is less wave energy released and also few submarine canyons to

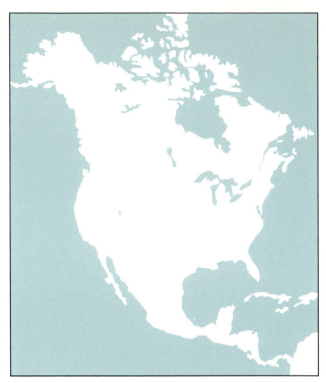

Figure 5.11 The coastline of North America

drain off sediment. Along this coastline then, there is sufficient sediment in the coastal zone to allow long, wide beaches to form.

The Gulf of Mexico is essentially a protected sea where both low-energy conditions and a low tidal range predominate. Moreover, the Mississippi River discharges approximately 300 million tons of sediment per year to these waters. As a result of these factors, the largest delta in the United States has formed, along with a chain of barrier islands that extends from the Florida panhandle to the Mississippi River Delta. The chain of islands divides this water body into the Mississippi Sound, which is between these barrier islands and the mainland, and the true Gulf of Mexico, which is offshore the islands.

Thus the coastlines of North America can be divided into glaciated and unglaciated regions that are composed of either consolidated or unconsolidated materials and into high-, moderate-, and low-energy environments on the basis of the energy released by breaking waves (fig. 5.11). All of these factors, as well as the presence or absence of submarine canyons, give these coastlines their distinctive features.

Summary

The sediments of the coastal zone may be either terrigenous or biogenous in origin. The terrigenous sediments originate from erosion of the adjacent landmasses or are carried into the sea by surface runoff or river flow. The biogenic sediments are either the shells or protective coverings of organisms that inhabit the adjacent coastal waters. Thus, coastal sediments consist of whatever materials are present on the adjacent landmasses suspended in river water or present offshore.

Once these sediments enter the marine environment, they are carried into the surf zone and are transported by the longshore current. When this current loses velocity, the sediments are deposited and form or augment beaches, mud flats, and so forth.

The coastlines of the north Pacific and Atlantic have been glaciated and often consist of durable bedrock. This material is very resistant to mechanical weathering; as a result, beaches are rare along these coasts. Where present, the beaches consist of pebbles or very coarse sand. Unglaciated coastlines consist of loosely packed sediment, which is moved about readily by waves and currents.

On the basis of the energy expended by breaking waves, coastlines may be considered to be high-, moderate-, and low-energy environments. The Pacific coast is a high-energy environment, where much sediment is moved about. A large portion of this sediment is siphoned off into submarine canyons. As a result beaches, where present, are narrow. The Atlantic coast is a moderate-energy environment and submarine canyons are not plentiful. Thus, these beaches are longer and broader. The Gulf coast, however, is a low-energy environment and very broad beaches have developed, along with the Mississippi River Delta.

Review

1. Distinguish between the foreshore and backshore on a beach.
2. How does wave action determine the slope of a beach?
3. Explain why summer beaches are composed of fine sand and generally have a gentle slope, while winter beaches are composed of coarser sand and have a steep slope.
4. What is the longshore current? How is it formed?
5. Distinguish between primary and secondary coastlines.
6. What are the major factors involved in moving sediment along or from a given coastline?
7. What accounts for the presence or absence of deltas?
8. What are the major differences between the Atlantic and Pacific coastlines?
9. What are the three types of coral reef? Describe the distinguishing features of each.
10. How does the presence or absence of submarine canyons affect the size and numbers of sandy beaches?

References

Bird, E. C. F. (1969). *Coasts.* Cambridge, Mass.: M.I.T. Press.

Johnson, D. W. (1946). *Shore Processes and Shoreline Development.* New York: Hafner Publishing Co.

King, C. A. M. (1960). *Beaches and Coasts.* New York: St. Martin's Press.

For Further Reading

Bascom, W. (August 1960). "Beaches." *Scientific American.*

Bascom, W. (1980). *Waves and Beaches.* New York: Doubleday.

Kaplan, E. H. (1982). *A Field Guide to Coral Reefs.* Boston: Houghton Mifflin Co.

6

Atoms, Molecules, and the Characteristics of Water

Key Terms

chemical bond
covalent bond
density
electron
first law of thermodynamics
hydrogen bonds
ionic bond
intermolecular attractive force
intramolecular attractive force

latent heat
long-range order
molecule
nonpolar covalent bond
polar covalent bond
proton
second law of thermodynamics
short-range order
specific heat

Everything in the universe is composed of extremely small particles called **atoms,** which are often bonded together to form molecules. Atoms consist of even smaller subatomic particles whose behavior gives atoms and molecules their characteristic properties.

Water is an unusual molecule with many unique chemical and physical properties. Perhaps the most distinctive property of water is its ability to form strong attractive forces. As a result of these attractive forces, large quantities of heat must either be removed from or released into the surroundings when water vaporizes, condenses, melts, or solidifies.

In addition, water is capable of storing very large quantities of heat, with only a small temperature rise in the water itself. For this reason water bodies warm up slowly in the spring and cool off slowly in the fall, thus preventing wide seasonal temperature fluctuations and moderating terrestrial temperatures in coastal areas.

Atoms and Molecules

From the standpoint of understanding the sea's chemistry, the most imortant of the atom's subatomic particles are the **electrons** and the **protons.** Protons are positively charged particles that are located in the central portion, or nucleus, of each atom. Electrons are negatively charged particles that orbit about the nucleus in electron clouds (fig. 6.1).

Since all atoms are electrically neutral and since the negative charge on an electron is equal in magnitude to the positive charge on a proton, each atom must have identical numbers of protons and electrons. Thus, the number of protons in the nucleus actually determines the number of electrons that will be present in the electron clouds. In essence it is the number of protons that gives each atom its characteristic properties, while the electrons participate in chemical reactions.

When substances react, electrons are either gained, lost, or shared by the atoms that participate in the reaction. The tendency of an atom to gain, lose, or share electrons is determined by the number of

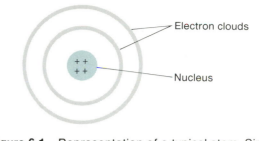

Figure 6.1 Representation of a typical atom. Since the nucleus in this example contains four positively charged protons, there must be four negatively charged electrons in the surrounding electron cloud.

protons within the atomic nucleus, as well as by the number of electrons that orbit about the nucleus. When two or more atoms react, their electrons are transferred or shared. This produces **chemical bonds** and results in the formation of **molecules.**

Chemical Reactions

Ordinary table salt is produced by the reaction of sodium and chlorine atoms. When a sodium atom encounters a chloride atom, the sodium loses an electron, which is captured by the chlorine. Before the reaction takes place, both of the atoms are electrically neutral, since they both have equal numbers of positively charged protons and negatively charged electrons. In the course of the reaction, the sodium atom loses a single unit of negative charge and becomes a positively charged particle, called an ion—in this case, a sodium ion. Conversely the chlorine gains this unit of negative charge and becomes a negatively charged particle, termed the chloride ion. Ions, therefore, are merely charged particles. Since the chemical symbol for sodium is Na and the symbol for chlorine is Cl, the reaction may be summarized as:

$$Na \longrightarrow Na^+ + 1\ electron$$
$$Cl + 1\ electron \longrightarrow Cl^-$$
$$Na^+ + Cl^- \longrightarrow {}^+NaCl^-$$

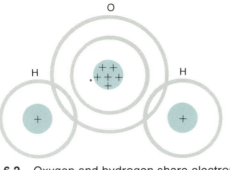

Figure 6.2 Oxygen and hydrogen share electrons to form the water molecule, which consists of two hydrogen atoms and a single oxygen atom.

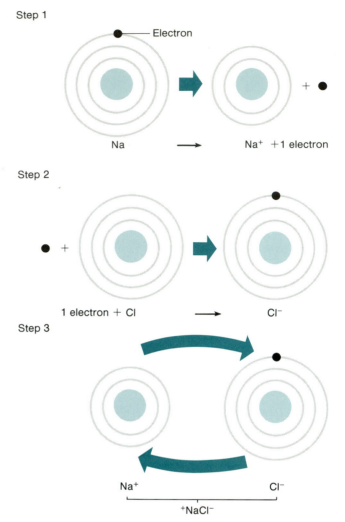

Figure 6.3 An ionic bond is formed as a result of the mutual attraction of positively and negatively charged particles. In this example sodium loses an electron and becomes positive (step 1). The electron is gained by chlorine, which becomes negative (step 2). The mutual attraction of the positive sodium and the negative chloride ion form the ionic bond (step 3).

It is the mutual attraction of the oppositely charged ions that actually forms the chemical bond, holds the Na and Cl together, and produces the sodium chloride molecule. Even after the molecule is formed, the sodium retains its positive charge and the chloride its negative charge. Thus, the molecule consists of fully charged positive and negative ends.

Water molecules are formed when molecular hydrogen gas, H_2, reacts with molecular oxygen gas, O_2. In the formation of water, however, electrons are neither gained nor lost but are, rather, unequally shared. In this case the oxygen attracts hydrogen's electrons, but not strongly enough to completely capture them. In addition each oxygen atom is capable of attracting two electrons, while each hydrogen atom has only a single electron to share. As a result each water molecule consists of one oxygen atom chemically bonded to two hydrogen atoms (fig. 6.2). This reaction may be summarized:

$$H_2 + O_2 \longrightarrow H_2O$$

As noted the atoms react in a 2:1 ratio. Two hydrogen molecules are needed in order to completely react with a single oxygen molecule. The water molecule is formed as a result of the unequal sharing of H and O electrons.

The formation of H_2O involves the reaction of H_2 with O_2, and the hydrogen and oxygen molecules consist of two identical atoms chemically bonded together. Since molecules of this type are composed of identical atoms, neither of these atoms has a greater tendency to attract electrons, and the electrons that form these chemical bonds are equally shared.

Chemical Bonds

Three types of electron-electron interactions may be involved in the formation of chemical bonds. When electrons are completely gained and lost, **ionic bonds** are formed (fig. 6.3) and the resultant molecule is termed an ionic molecule. When, on the other hand, the atoms that compose a molecule share electrons, these atoms are held by **covalent bonds.**

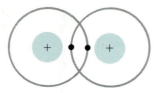

Figure 6.4 When a molecule consists of identical atoms, neither exerts a greater or lesser attraction. The electrons are equally shared and the atoms are held by a nonpolar covalent bond.

Figure 6.5 Although the oxygen atom is unable to completely capture the hydrogen electrons, it does attract them rather strongly. As a result the electrons are unequally shared and spend a greater amount of time in the vicinity of the oxygen atom. The oxygen attains a partial negative charge, while each hydrogen develops a partially positive charge.

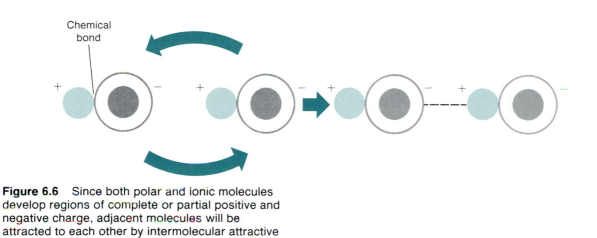

Figure 6.6 Since both polar and ionic molecules develop regions of complete or partial positive and negative charge, adjacent molecules will be attracted to each other by intermolecular attractive forces.

Since H_2 and O_2 consist of identical atoms, the electrons must be equally shared. As a result there are no negative and positive regions. Molecules of this type are known as nonpolar covalent molecules and their chemical bonds are called **nonpolar covalent bonds** (fig. 6.4).

In the case of the water molecule, however, the electrons are unequally shared, and each hydrogen electron spends more time in the vicinity of the oxygen atom. The oxygen atom therefore develops a partial negative charge ($\propto -$), while each hydrogen atom develops a partial positive charge ($\propto +$). Molecules having partially positive and negative regions are polar covalent molecules and the chemical bonds that hold them together are called **polar covalent bonds** (fig. 6.5).

The Liquid, Solid, and Vapor State

Since ionic and polar covalent molecules have charged regions, attractions develop between the positively charged portion of one molecule and the negative areas of adjacent molecules (fig. 6.6). These forces are termed **intermolecular attractive forces** to distinguish them from the stronger chemical bonds, which may be thought of as **intramolecular forces.** The extent and strength of these intermolecular attractive forces are dependent upon the type of electron transfer that occurs between the atoms that compose the molecule.

Box 6.1   The Laws of Thermodynamics

The laws of thermodynamics describe the behavior of both matter and energy. The first and second laws are of particular interest in explaining and interpreting the behavior of atoms and molecules. The **first law of thermodynamics** states that neither matter nor energy is created or destroyed but can be, and often is, converted from one form into another. When atoms and/or molecules react, they invariably form molecules that have characteristics very different from the reactants. For example, hydrogen reacts with oxygen to form water. Both H_2 and O_2 are gases at normal earth temperatures, yet the resultant molecule is a liquid at the same temperature. In this reaction neither the hydrogen nor the oxygen disappears but is converted into a different chemical form.

The formation of water and, indeed, all chemical reactions are examples of one of the aspects of the first law of thermodynamics, since matter was neither created nor destroyed but was converted from one form to another.

The first law of thermodynamics also considers energy changes. Energy is present in the universe in many diverse forms, such as the radiant energy of the sun, readily available and highly concentrated

chemical energy, and the more randomly dispersed heat energy, which, since it is dispersed, is less readily available for useful purposes. All of these forms of energy are convertible from one form to another, and the total amount of energy present in the universe always remains constant. For example, when coal is burned, a large amount of energy is released in the form of heat. This energy, prior to the combustion of the coal, was present in the form of chemical bond energy that served to hold together the molecules that composed the coal. In this example highly structured matter and concentrated, high-quality chemical energy was converted to a less organized, more highly dispersed mixture of gases and low-quality heat energy. Since this energy is dispersed throughout the surroundings, rather than concentrated in the coal, the products are at a lower energy state than the reactants.

In the course of this reaction, a very small amount of coal is converted to energy. Matter and energy are therefore convertible and intimately related to each other. As a consequence any attempt to deal with one must involve the other. Einstein devised the famous equation $E = mc^2$ to illustrate this relationship and to show that matter and energy are interchangeable. In this equation E

Ionic compounds such as sodium chloride have totally positive and negative regions. As a result strong attractive forces develop between the positive region of one molecule and the negative region of an adjacent molecule. These attractions serve to hold large aggregations of ionic molecules together in a highly structured crystalline lattice (fig. 6.7). Due to these strong attractive forces, ionic compounds are said to have **long-range order,** which enables these compounds to exist in the solid state. Solids have a definite shape and occupy a definite volume, which is the direct result of the molecules' adhering rigidly to each other.

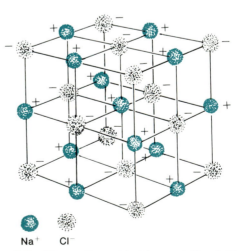

Na$^+$ Cl$^-$

Figure 6.7   The sodium chloride crystalline lattice results from the strong intermolecular attractive forces that develop between adjacent molecules.

represents energy; *m,* mass; and *c,* the velocity of light, light being a form of energy. The first law of thermodynamics can be most accurately stated: Matter and energy can neither be created nor destroyed, but, since they are interchangeable, they can be converted from one form to another.

In the course of chemical reactions, matter and energy are changed into different forms. When atoms react to form molecules, the same number of atoms that were present in the reactants must, according to the first law of thermodynamics, be present in the products. As in the burning of coal, energy changes commonly accompany these chemical changes. When chemists balance equations, it is merely a form of bookkeeping dictated by the first law. As a result the equations used to describe the previous reactions are more accurately written:

$$2Na + Cl_2 \longrightarrow 2NaCl$$
$$2H_2 + O_2 \longrightarrow 2H_2O$$

In the formation of NaCl, solid sodium reacts with gaseous chlorine to form solid table salt, and, as discussed, water is produced by the reaction of two gases, oxygen and hydrogen. As the balanced equations indicate, neither hydrogen, oxygen, sodium, nor chlorine is destroyed, nor is matter created to form the products. The reactants are merely rearranged to form the products that contain identical numbers of atoms. Energy changes accompany these chemical changes, and the energy of the chemical products is lower than the energy of the reactants. During these reactions it is actually the movement of electrons among the atoms that form the chemical bonds that hold these molecules together. This movement of electrons between atoms and the tendency of some atoms to gain, others to lose, and yet other atoms to share electrons, involve changes in energy of atoms and are governed by the **second law of thermodynamics.**

Essentially the second law of thermodynamics states that all matter and energy will tend to become less structured, and therefore energy tends to become less available for useful purposes. When any type of matter chemically decomposes, its highly structured molecules are broken down into a mixture of smaller atoms and molecules, and a portion of the chemical bond energy that served to impose order on the matter becomes dispersed to the surroundings. The result is that the molecules that originally composed the matter attain a lower energy state. Indeed all matter, over the long term, tends to react spontaneously in such a manner as to attain its lowest possible energy state.

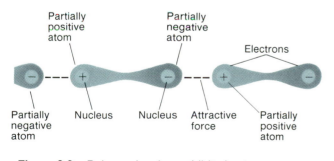

Figure 6.8 Polar molecules exhibit short-range order, since they do not form large aggregations.

Polar covalent molecules such as water, on the other hand, have only partially positive and partially negative regions. Although the positive region of one molecule is attracted to, and attracts, the negative region of an adjacent molecule (fig. 6.8), the attractive forces are not as great as those developed between ionic molecules, since the charges that are present on polar molecules are only partial. As a consequence polar molecules do not form very large aggregations in highly structured arrangements. For this reason, polar covalent molecules are said to have a **short-range order** and therefore exist primarily in the liquid state. Since these molecules adhere firmly, but not rigidly, liquids are characterized by their ability to flow. If the intermolecular attractions are increased, the tendency to flow decreases. Viscosity is a measure of a substance's resistance to flow. In this case the viscosity would increase. The viscosity of water determines the rate at which objects will sink, as well as the rate at which water will move. All liquids have a definite volume but not a definite shape.

Gases consist of either single atoms, such as helium and neon, or nonpolar molecules, such as H_2, O_2, Cl_2, and so forth. Since they are composed of either single atoms or identical atoms that are chemically bonded together, the atoms and/or molecules that comprise a given gas do not have positive and negative regions, and attractive forces, if they form at all, are transitory and insignificant. The result is that gases have no order and are characterized by having no specific shape, since the particles move totally independently of each other.

In order to change a solid to its liquid state or a liquid to its gaseous state, the molecular order of the substance must be disrupted by decreasing the attractive forces. Conversely, to change a vapor to its liquid state or a liquid to its solid state, the molecular order must be increased by allowing additional attractive forces to develop. This is most easily accomplished by heating and cooling the substance.

Heat and Changes of State

Temperature changes accompany the change in state of any compound. For example, in order for any liquid to enter the vapor state, sufficient energy in the form of heat must be added to convert the liquid into its gaseous form. The heat serves to increase the motion of the molecules that compose the liquid, which in turn disrupts the intermolecular attractive forces. Eventually some of the molecules attain sufficient energy, break through the surface of the liquid, and enter the vapor state. The heat energy has been transferred from the surroundings; when these molecules break through the surface as a vapor, they have a greater energy than they had as a liquid. This phenomenon is important in the sea, since when water evaporates it removes heat from sea water and transfers it to the atmosphere as high-temperature water vapor.

In order for the water or any vapor to return to its liquid state, energy in the form of heat must be removed from the vapor. This loss of energy reduces the speed of the molecules that comprise the vapor. As the molecules lose speed, intermolecular attractive forces begin to develop and further restrict the molecular motion. This process enables a greater number of molecules to occupy a given volume. Eventually these molecules will condense and return to the liquid state.

In order for a liquid to solidify and enter the solid state, additional energy must be removed. This loss of energy in the form of heat further reduces molecular motion and increases intermolecular attractions. At a specific temperature, the molecular motion becomes so reduced and the intermolecular attractions so great that the liquid enters the solid state. The temperature at which this occurs is the freezing point of that specific liquid. When pure water freezes, the temperature will remain constant at $0°C$ ($32°F$) until all of the liquid has entered the solid state. The temperature remains constant at the freezing point, since heat is released to the surroundings as each molecule enters the solid state. Once the system is completely frozen, the temperature of the ice will fall to the surrounding atmospheric temperature.

To reverse the process and melt the solid, heat must be added. This addition of energy increases the molecular motion and disrupts the intermolecular attractive forces. Energy is transferred to the molecules, and the solid melts.

The Water Molecule and Hydrogen Bonds

When hydrogen reacts with oxygen to form water, the hydrogen electron is unequally shared, rather than completely lost to the oxygen atom. Since the hydrogen electron spends a greater amount of time in the vicinity of the oxygen atom, the hydrogen atom assumes a partially positive charge, while the oxygen atom develops a partially negative charge. These charges, though partial, are strong due to the high electron affinity of the oxygen atom.

Thus, the water molecule consists of two partially positive ends and a partially negative middle region. These regions of opposite charge are mutually attracted to each other, and very strong chemical bonds form between the hydrogen and oxygen

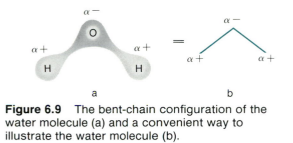

Figure 6.9 The bent-chain configuration of the water molecule (a) and a convenient way to illustrate the water molecule (b).

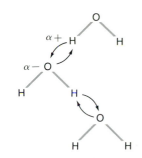

Figure 6.10 Due to the development of the partial charges on the atoms, the hydrogen and oxygen atoms on adjacent molecules are attracted to each other.

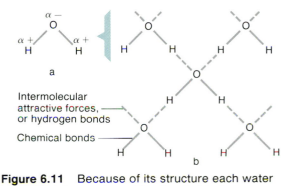

Intermolecular attractive forces, or hydrogen bonds ——→

Chemical bonds ———

Figure 6.11 Because of its structure each water molecule (a) attracts and is attracted to four adjacent water molecules (b).

atoms. In order to achieve maximum spacing between the electrons that orbit about the atoms, water molecules assume a "bent-chain" configuration, with each hydrogen atom bonded to the oxygen atom at an angle (fig. 6.9).

Due to its polarity, each water molecule in both the liquid and solid state orients itself in a predictable manner in relation to its surrounding water molecules. The partially negative oxygen atom of one water molecule is attracted to and attracts the partially positive hydrogen atoms of adjacent molecules (fig. 6.10). As a result of water's molecular configuration, intermolecular attractive forces are able to develop between each water molecule and four adjacent water molecules (fig. 6.11) so that each attracts and is attracted to four other molecules. This maximizes the intermolecular attractive forces that develop in water. In addition the intensity of the charges on the atoms enables intermolecular forces to form between water molecules that are considerably stronger than those that form between other polar compounds. Hence the intermolecular attractive forces that develop in water are numerous and strong. These bonds are called **hydrogen bonds** to distinguish them from other, weaker intermolecular attractive forces.

Hydrogen Bonding and State Changes

Because of the presence of hydrogen bonds, water has remarkably high freezing and boiling points. The boiling point of pure fresh water is 100°C (212°F); its freezing point is 0°C (32°F). Based on the properties of nonpolar compounds that have molecular structures similar to water, it has been calculated that if water were nonpolar, it would boil at −80°C (−176°F) and freeze at −100°C (−212°F). In

other words, if water were nonpolar, ice and liquid water would not exist, and water would be a vapor at the temperatures commonly found on earth.

In order for a water molecule to enter the vapor state, the hydrogen bonds that hold it to adjacent molecules must be broken. Following this the molecule must then gain sufficient energy in the form of molecular motion to enable it to break through the surface of the liquid (fig. 6.12). Since the hydrogen bonds are both numerous and strong, a considerable amount of energy is required in order to vaporize water.

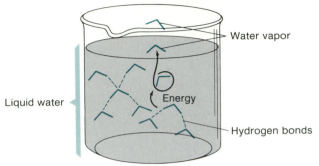

Figure 6.12 In order to evaporate water, the hydrogen bonds must first be broken. Then additional energy must be added to enable the water molecules to break through the surface of the liquid and enter the vapor state.

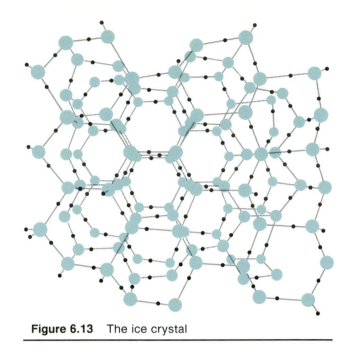

Figure 6.13 The ice crystal

By comparison nonpolar and slightly polar compounds do not form bonds of this magnitude and strength; each molecule moves about independently, except when intermolecular attractions occur. As noted previously, these attractive forces are considerably weaker than hydrogen bonds and are also transitory. Consequently little heat is needed to disrupt these attractive forces and allow the molecules to move about as independent particles. Little additional energy is then required to enable these molecules to break through the surface of the liquid and enter the vapor state. The boiling points of nonpolar and slightly polar liquids are therefore considerably lower than a highly polar liquid such as water.

Water also freezes at a relatively high temperature as compared with other liquids. This too is due to its tendency to form hydrogen bonds. As liquid water is cooled, heat is removed from the liquid, which reduces the molecular motion and enables additional hydrogen bonds to form. As a result of their polarity, water molecules are readily attracted to each other, facilitating the formation of ice crystals (fig. 6.13).

Nonpolar and slightly polar compounds enter the solid state at much lower temperatures. This is also due to the absence of hydrogen bonds. Since they are nonpolar or only slightly polar, there is little to attract the molecules to each other and enable them to enter the solid state. Large amounts of heat must be removed in order to reduce molecular motion to the point at which the molecules are moving slowly enough to enable the intermolecular attractive forces to become effective, thereby enabling the molecules to enter their solid state.

Specific and Latent Heat of Water

Hydrogen bonds must be formed or broken to change the state of water. In order to melt ice and evaporate liquid water, heat is required to increase the motion of the molecules and break the hydrogen bonds that give the solid its long-range order and the liquid its short-range order. Conversely heat must be removed to condense water vapor and freeze liquid water. The removal of heat results in a decrease in molecular motion, enabling a greater number of hydrogen bonds to form. Increased hydrogen bond formation increases the molecular order that characterizes the liquid and solid states.

Since large amounts of heat energy are involved in the formation and breaking of hydrogen bonds, only small temperature changes accompany the transfer of heat to a body of water. Water is therefore capable of storing very large quantities of heat and is said to have a high **specific heat.**

Specific heat is defined as the amount of energy that is required in order to raise the temperature of one gram of a substance 1°C (1.8°F). This energy is generally measured in units called calories. In fact water is used as the standard measurement for a calorie; a calorie is defined as the amount of energy that is required to raise the temperature of one gram of water 1°C. Based on the definition of specific heat, therefore, water must have a specific heat of 1, which is the fourth highest specific heat of all known compounds.

The high specific heat of water is responsible for its high heats of fusion and evaporation. These terms refer respectively to the number of calories that must be removed in order to freeze water and the number of calories that are required to evaporate it. The heat of fusion and heat of evaporation involve changes in state and may be conveniently discussed in terms of **latent heat.**

Latent heat is defined as the number of calories needed to change the state of a substance at a given temperature and pressure. Since water has the highest heat of fusion and evaporation of any naturally occurring liquid, it also has a very high latent heat.

The addition or removal of heat affects the motion of water molecules, breaks or forms hydrogen bonds, and enables the water to occupy a greater or lesser volume. The increase or decrease in molecular motion and the subsequent formation or disruption of the hydrogen bonds enables water to store large quantities of energy with a minimal temperature change, giving water its high specific heat. The high specific heat, in turn, is responsible for the large amounts of energy that must be added in order to evaporate water or must be removed to form ice.

All these factors are involved in giving water many of its unique properties. Temperature effects also cause water masses to layer or stratify, while the high specific heat and latent heat of water play major roles in climatic patterns.

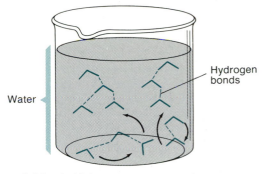

Figure 6.14 At higher temperatures there are fewer hydrogen bonds and the molecules are able to move about more freely—they occupy a greater volume.

Temperature, Hydrogen Bonds, and Density

As the temperature of water is increased, the molecular motion will increase, causing hydrogen bonds to break. The water molecules will then be able to move about more freely and occupy a greater volume. The reverse occurs when water is cooled. Since cool water contains more molecules in a given volume than warm water, it can be considered to be heavier than warmer water. To be precise, however, the water is said to be denser. The concept of **density** is used to relate the mass of a substance to the volume occupied by that substance and may be summarized by the equation Density = mass/volume, or $D = m/v$.

If the water volume is increased while the mass of the water remains the same, the density of the water must decrease. Conversely if the volume is decreased and the mass kept constant, the density must increase, since the same number of water molecules would occupy a smaller volume. Assume, for example, that there is 1 kilogram of water in a beaker and that this water occupies a volume of 1 liter (fig. 6.14). Using the equation $D = m/v$, the density of the

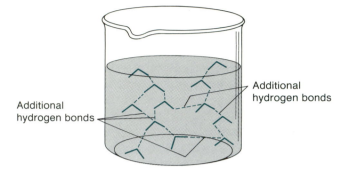

Figure 6.15 As water is cooled the molecules lose energy and slow down, and additional hydrogen bonds form. This restricts the motion of the molecules. They occupy less space, and the volume is reduced.

water would be D=1 kilogram/1 liter, or D=1. If this water is cooled, the water molecules will slow down, allowing more hydrogen bonds to form. The formation of additional hydrogen bonds will further restrict the motion of the water molecules, and they will occupy less space (fig. 6.15); that is, the volume occupied by the water will decrease. Since no water has been added to or removed from the beaker, neither the number of water molecules nor the weight of the water has changed; only the volume occupied by the water has changed. If the volume occupied by the water at this lower temperature is only 0.75 liters, then the density would be 1 kilogram/0.75 liters or D=1.33. The water did not become heavier but it did become denser, since the same number of molecules are packed into a smaller space. However, the weight of the molecules per unit volume did increase. In other words, 100 milliliters of water at 10°C (50°F) would contain more water molecules than 100 milliliters of water at 50°C (122°F). Since there are more water molecules per unit volume at 10°C, this water would be denser and would sink beneath the warmer water.

If, on the other hand, the water is warmed, the molecular motion will increase, hydrogen bonds will break, and the molecules will be able to occupy a

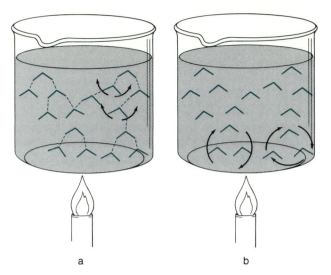

Figure 6.16 As the energy is increased by heating, hydrogen bonds will break (a). As the number of hydrogen bonds are reduced (b), the molecules will be present in smaller aggregations and will be able to move about more freely. They will occupy more space, and the volume will increase.

greater volume (fig. 6.16). For example, if the 1 kilogram of water occupies a volume of 1.2 liters at the higher temperature, its density would be 1 kilogram/1.2 liters or D=0.83. In this instance the water did not really become lighter, since there is still 1 kilogram of water in the beaker, but the molecules have "spread out" and now occupy a greater volume. Since there are fewer molecules per unit volume at the higher temperature, the weight per unit volume did decrease. Rather than speaking of weight changes in relation to temperature, however, it is customary to speak of density changes.

In either case, and with the important exception of very cold liquid water, a decrease in temperature will increase the density of water and an increase in temperature will decrease the density. As a result, cold dense water will sink beneath warmer, less dense water.

Density and the Liquid, Solid, and Vapor States of Water

If one gram of water were placed into a one-liter container, the container sealed so that no water could enter or escape, and the temperature raised to and maintained at 150°C (302°F), the water would enter the vapor state and all the water molecules would move about independently. If two of these molecules did happen to encounter each other, their

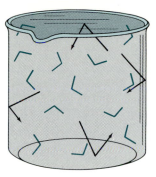

Figure 6.17 When water is maintained at a temperature above its boiling point, all of the hydrogen bonds are broken, and each water molecule moves about independently—the water is in its vapor state. When the molecules strike the walls of the container, they exert pressure. Note that the water occupies the entire container in this example.

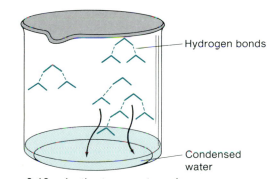

Hydrogen bonds

Condensed water

Figure 6.18 As the temperature decreases, hydrogen bonds form; the molecules form into aggregations and the larger aggregates condense and fall to the bottom as liquid water.

speed would be too great for hydrogen bonds to be able to form. As the molecules moved about, they would strike and exert pressure on the walls of the container. This is termed vapor pressure, which if measured at any point, would be the same. Since the pressure exerted by the molecules is equal at all points, it can be said that the water vapor is effectively filling the entire container (fig. 6.17), and the volume occupied by the 1 gram of water would be 1 liter. In this case, therefore, the density of the water would be 0.001 gram/milliliter or D=0.001.

If the container were then cooled, the water molecules would begin to lose speed, and hydrogen bonds would begin to form when they encountered other water molecules at this lower speed. Eventually several water molecules would become linked together by hydrogen bonds (fig. 6.18), and the water would condense.

If the container were cooled to 50°C (122°F), essentially all of the water molecules would have returned to the liquid state and each would be hydrogen bonded to at least one other water molecule (fig. 6.19). Since 1 gram of water occupies a volume of 1 milliliter, when this sample is completely condensed, it would occupy only 1 milliliter of the container's 1 liter volume. Hence, the density would be 1.0 gram/milliliter or D=1.

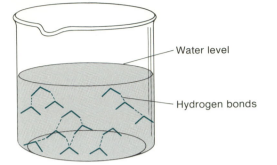

Figure 6.19 At 50°C. (122°F) all of the water molecules have condensed. Note that the water is not occupying the entire container at this point. The volume occupied by the water has decreased considerably.

Figure 6.21 At 4°C. (39°F) there is maximum hydrogen bond formation. The water molecules occupy the minimum volume and thus are at their maximum density.

Figure 6.20 As the temperature continues to decrease, additional hydrogen bonds are formed. This causes a greater restriction in molecular motion and an additional reduction in the volume of the water.

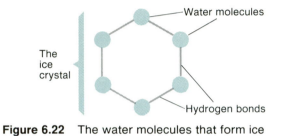

Figure 6.22 The water molecules that form ice crystals are widely spaced. As a result they occupy a greater volume and are less dense.

If the container continued to cool, the water molecules would continue to lose speed. Additional hydrogen bond formation would further restrict molecular motion (fig. 6.20), causing an additional decrease in the volume occupied, and the water would get denser.

As the water continued to cool, more and more hydrogen bonds would form. When the temperature reached 4°C (39°F), there would be maximum hydrogen bond formation. At this temperature the water molecules have an extremely limited motion and, as

a consequence, are so closely packed that they would occupy the minimum possible volume (fig. 6.21). Thus, fresh water is at its maximum density at 4°C.

When water reaches 0°C (32°F), it will freeze, become less dense, and float in the liquid water. This is due to the structure of the ice crystal (fig. 6.22), which consists of six water molecules connected by hydrogen bonds in a three-dimensional hexagonal crystalline structure. As a result of the fixed directions of the hydrogen bonds, the structure of the ice crystal is an open network of water molecules. There is a great deal of space between each water molecule; consequently, the entire ice crystal occupies a rather large volume. This spatial arrangement decreases the density of ice and causes it to float in the liquid water.

The density does not suddenly decrease when the water molecules reach 0°C, however. Rather the density begins to decrease as the water cools below 4°C. At temperatures below 4°C, but higher than

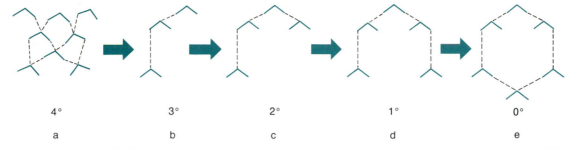

4°	3°	2°	1°	0°
a	b	c	d	e

Figure 6.23 Water at 4°C (39°F) is at its maximum density (a). As the temperature is decreased (b), some of the molecules are cooled to 3°C (37°F) and begin to spread out prior to entering the true crystalline structure. The volume, therefore, increases and the density decreases. Thus, the 3°C water is less dense and will float above the 4°C water. Should cooling continue (c) some of the molecules will reach 2°C (35.5°F), the volume will increase, the density will decrease, and the 2°C water will float above the 3°C water, which is located above the 4°C water. Similarly, when some of the water (d) reaches 1°C (34°F), it will be less dense and will float above the warmer water (a,b,c). Eventually ice forms (e), which has an even lower density and floats above liquid water of all other temperatures.

0°C, the water molecules begin to move apart (fig. 6.23) prior to entering into the true crystalline structure. Water at these temperatures is termed pseudocrystalline water. This arrangement decreases the density of the liquid water as the temperature falls below 4°C and causes it to float over the denser water. As the pseudocrystalline water continues to cool, additional water molecules enter the configuration. For example, molecules at 2°C (36°F) will float over the 3°C (37°F) water which, in turn, is located above the water that is at 4°C (fig. 6.24).

Thus, as water vapor or liquid water is cooled to 4°C, the reduction in temperature decreases the molecular motion which, in turn, facilitates hydrogen bond formation, resulting in a decrease in volume and an increase in density. At 4°C there is maximum hydrogen bond formation, and the molecules occupy the minimum volume. The water density is greatest at this temperature. Below 4°C the water molecules begin to spread out prior to forming the ice crystal. This rearrangement results in the pseudocrystalline water's becoming less dense and floating above the relatively warmer, denser water. At 0°C, the water molecules form the open, hexagonal ice crystal. Since these molecules occupy a greater volume, ice is less dense and floats.

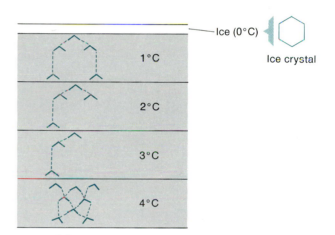

Ice (0°C)

Ice crystal

1°C

2°C

3°C

4°C

Figure 6.24 Water will tend to layer as a result of the different densities at different temperatures. The 4°C (39°F) water will always be on the bottom, since it is the densest.

Summary

Molecules are formed as the result of the transfer of electrons between atoms. The complete loss and gain of electrons results in the formation of ionic molecules, which have completely positive and negative regions. Unequal sharing of electrons, on the other hand, characterizes the polar covalent molecules, which have partially positive and negative regions. The equal sharing of electrons results in the formation of nonpolar covalent molecules, which do not develop charged regions.

Due to the development of charges on ionic and polar molecules, intermolecular attractive forces form between these molecules and enable the compounds to exist in the solid and liquid state. Ionic compounds have long-range order and exist as solids. Polar covalent molecules are liquids because of their short-range order, while the nonpolar gases do not develop intermolecular attractions and, as a result, exhibit no order.

Changes in state are due to a change in the order of compounds. When energy is added, molecular motion increases and intermolecular attractive forces are disrupted. This results in the melting of solids and the evaporation of liquids. When energy is removed, the molecular motion is decreased, which increases the formation of intermolecular attractive forces. This enables vapors to condense and liquids to freeze.

The structure of the water molecule allows it to readily form very strong intermolecular attractive forces, termed hydrogen bonds, with four adjacent water molecules. These hydrogen bonds give water its high boiling and freezing points, as well as its high specific and latent heats. As a result, water can store large quantities of energy with only a small temperature change.

Review

1. Draw an atom with three protons in its nucleus.
2. What are the differences between chemical and hydrogen bonds?
3. What are the differences between ionic, polar covalent, and nonpolar covalent molecules?
4. Compare and contrast intermolecular and intramolecular attractive forces.
5. Why is water a polar molecule?
6. How does the polarity of water affect its properties?
7. Why are ionic compounds generally solids?
8. Why is hydrogen (H_2) a gas?
9. What are long- and short-range order? How does each come about?
10. Explain why ice floats.

References

Black, J. A. (1977). *Water Pollution Technology.*
    Virginia: Reston Publishing Co.
Hein, M. (1984). *Foundations of College Chemistry.*
    California: Wadsworth.

For Further Reading

Masterton, W. L., Slowinski, E., and Stanitski, C. (1983).
    Chemical Principles. New Jersey: Saunders.
Slenko, M. J. and Plane, R. A. (1976). *Chemistry.* New
    York: McGraw Hill.

7

Water and the Waters of the Sea

Key Terms

aphotic zone
biological magnification
chemical availability
density barrier
density differential
dysphotic zone
dissociation
epilimnion
euphotic zone
homothermous

hypolimnion
juvenile water
polychromatic light
pressure-gradient current
salinity
salinity density barrier
spatial unavailability
temperature density barrier
thermocline

Water in its liquid state is profoundly affected by temperature changes. As the temperature increases, the intermolecular attractive forces are disrupted, allowing the molecules to move about more readily and occupy a greater volume. When the temperature is decreased, the reverse occurs and the molecules occupy less volume. There are, as a result, more water molecules per unit volume in cold water than there are in warm water. Also the cooler water will sink beneath the warmer water. This phenomenon is a major factor in the formation of currents in marine systems.

Since water is polar, it is able to dissolve virtually every element known to man and is often referred to as the universal solvent. These dissolved materials give seawater its characteristic salinity. Salinity in a water mass increases its density; thus, a water mass of higher salinity will sink beneath one of lower salinity. Salinity—in this case the salinity differences between adjacent water bodies—is also a major factor in the formation of currents in the deepest ocean and the shallowest bay.

Some materials dissolve more readily in substances other than water; as a result, only small amounts of these materials are present in a water body at any given time. Since such materials are often easily dissolved in fatty substances, they may be removed from the water by marine organisms and accumulate in those organisms. This is one of the major factors involved in the mercury contamination of tuna fish. This same characteristic is also the explanation for such diverse occurrences as the contamination of plants and animals by pesticides, trace metals, and other such substances and for the thinning of eagles' egg shells.

Temperature Effects on Natural Systems

To truly appreciate the effects of temperature on hydrogen bonding and the effects of hydrogen bonding on density, it is necessary to consider a large natural system. Deep lakes in the temperate zone provide excellent examples, since salinity interactions do not occur in these freshwater systems.

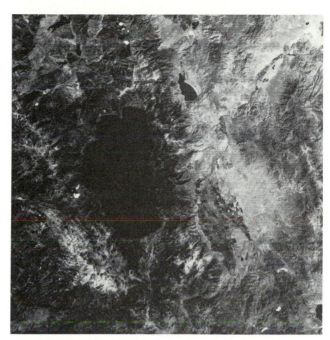

Temperature density barriers commonly form in Lake Tahoe.

If the temperature of such a lake were measured in the early spring, the entire lake would be at 4°C (39°F). Systems that are at a uniform temperature are said to be **homothermous**. As the season progresses, the sun's rays strike the lake more directly, which, in conjunction with warmer air temperatures, causes the surface waters to warm above 4°C. As a result the molecular motion increases, hydrogen bonds begin to break, the molecules occupy a greater volume, the density decreases, and the warmer water remains on the surface. A portion of the heat is transmitted to the subsurface waters, causing them to increase in temperature, become less dense, and layer above the denser water.

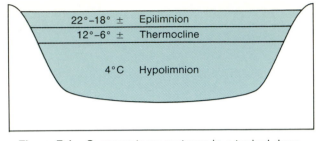

Figure 7.1 Summer temperatures in a typical deep lake in the temperate zone

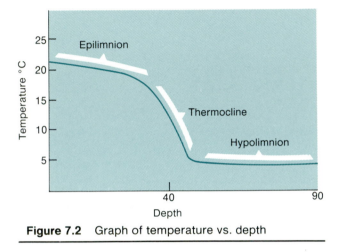

Figure 7.2 Graph of temperature vs. depth

By midsummer a definite temperature stratification can be observed (fig. 7.1). The colder, denser water will be beneath the warmer, less dense water. If the temperature is plotted against depth (fig. 7.2), a gradual temperature decline in the surface waters is observed. This area of gradual temperature change is called the **epilimnion.** Beneath the epilimnion is a zone of rapid temperature change, known as the **thermocline.** In order for a layer of water to be considered a thermocline, the water temperature must change by at least 1°C for each meter in depth. The waters of the thermocline lie directly above the densest bottom water. In this region, called the **hypolimnion,** the water is at a constant temperature of 4°C.

The considerable temperature differences between the surface and the bottom waters result in a large **density differential** between these water masses. The thermocline acts as a **density barrier** and effectively prevents the epilimnion from mixing with the hypolimnion. Since this density barrier is based solely on temperature effects in a freshwater lake, it is termed a **temperature density barrier.**

In the autumn such a lake will gradually cool and become homothermous at 4°C, at which point the temperature density barrier will disappear. During the onset of winter, the surface waters will cool below 4°C, causing the temperature density barrier to re-form. This time, however, the colder, less dense water will be in the epilimnion and the warmer 4°C water will be found in the hypolimnion. In both winter and summer, the dense 4°C water will always be at the bottom of the lake, separated from the less dense surface water by a temperature density barrier.

In the sea the materials dissolved in the water will increase its mass and, thereby, increase its density. As a result, temperature effects are augmented by the concentration of dissolved materials in marine systems.

Heat and the Sea

With the exception of the small amount of heat that enters the sea from the earth's interior, the vast majority of heat is derived from the sun. This heating is not uniform, however; it is at its greatest when the sun is more directly overhead. The sun is in this position during the summer months in the temperate areas of the Northern Hemisphere and almost continually in the tropics and the equatorial regions.

Despite the large quantities of heat that enter the seas in the tropics and at the equator, these water areas are not getting warmer than those of the temperate and polar regions, nor is the ocean, in total, becoming warmer. The energy of the tropical areas is transferred to both the atmosphere and to the more northerly and southerly waters. The major factors that serve to transfer this energy are evaporation, back radiation, contact conduction, winds, and water currents.

Evaporation, as discussed previously, is the conversion of high-energy liquid water to the vapor state; back radiation is merely the radiation of heat from the sea back to the atmosphere; and contact conduction is the transfer of heat directly from the sea's surface to the overlying air. Evaporation, back radiation, and contact conduction ultimately form winds that produce water movements known as wind-drift currents (chapter 10). In all of the world's ocean, evaporation removes 51 percent of the heat; back radiation, 41 percent; and contact conduction, 7 percent. There is, however, a net gain of heat at the equator and a net loss in the temperate and polar regions. The winds and wind-drift currents serve to redistribute this heat and maintain the long-term constant temperature regime.

When water evaporates, its molecular motion is increased to the point at which all of the hydrogen bonds are broken and the water molecules attain sufficient energy to break through the surface of the water and enter the vapor state. This high-energy water vapor, along with the air warmed by back radiation and contact conduction, is warmer than, and therefore less dense than, the air above it. As a result the warmer air and water vapor rise into the upper atmosphere.

As the water vapor rises, it cools. Hydrogen bonds form and the water condenses and ultimately returns to earth as precipitation, primarily in the equatorial regions. The air masses also cool, increase in density, and descend through the lower, warmer, less dense air masses. These descents generally occur in the tropical and subtropical areas. As the descending air nears the earth's surface, it begins to move horizontally and set surface waters in motion as wind-drift currents. The Gulf Current (chapter 11) is a well-known wind-drift current that moves warm equatorial waters northward in the Atlantic Ocean.

Once the wind sets the water in motion, the transfer of heat between the moving air and water masses is facilitated. The heat is transferred by contact conduction, and the direction of heat flow is determined by the temperature differentials—temperature gradients—between the water and the overlying air mass. If the air is warmer than the sea, heat will be transferred from the atmosphere to the sea. If the water is warmer than the air, heat will be transferred to the air mass. Since wind-drift currents tend to move warm water into the cooler northern and southern latitudes, heat tends to leave the water and enters and warms the cool overlying air. The result is that the wind-drift currents and air–sea temperature gradients both maintain the stable temperatures that are characteristic of the world's seas and moderate air temperatures on a global scale. Air–sea interactions are discussed in chapter 10.

Figure 7.3 The partially negative oxygen atoms of the water molecules are attracted to and attract the positively charged sodium atoms, while the partially positive hydrogen atoms attract and are attracted to the negatively charged chlorine.

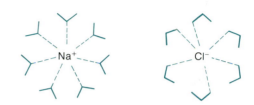

Figure 7.4 The dissociation of sodium chloride into positively charged sodium ions and negatively charged chloride ions

In essence these phenomena are due to the energy required to break the hydrogen bonds between adjacent water molecules. Since hydrogen bond formation is due to the polarity of the water molecules, these occurrences are ultimately the result of the physical structure of water.

The polarity of water is also responsible for its ability to dissolve virtually every known element. This gives seawater its characteristic concentration of dissolved materials, which profoundly affects the movement and location of water masses.

Water: The Universal Solvent

When an ionic molecule such as sodium chloride is added to water, the water molecules, because of their polarity, arrange themselves about the NaCl in a predictable manner. The partially negative oxygen atoms are attracted to and gather about the positive sodium, while the negative chlorine is surrounded by the partially charged hydrogen atoms (fig. 7.3). Intermolecular attractive forces develop, and the NaCl molecule separates, or **dissociates,** into a negatively charged chloride ion and a positively charged sodium ion (fig. 7.4). When polar covalent mole-

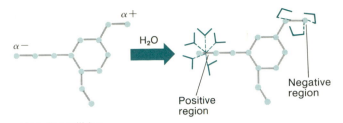

Figure 7.5 Complex polar molecules will be surrounded by water molecules and will dissolve. Since they are polar rather than ionic, they will not dissociate.

An excessive depletion of oxygen often causes extensive fish kills.

cules are added to water, the same interactions occur to a lesser degree. Large polar molecules, such as many carbohydrates and proteins, dissolve, but do not dissociate in water (fig. 7.5).

In these examples the water is termed the solvent, the dissolved materials are known as the solute, and the entire system is referred to as a solution. Generally the solvent is defined as the material that is present in the larger amount; the solute, the material present in the smaller amount. Since virtually every compound is soluble in water to some extent, water is generally considered to be the solvent in any solution in which it is present and thus is often called the universal solvent.

Even nonpolar material such as oxygen dissolves to a slight degree in water. Water's ability to dissolve oxygen is due to the high degree of polarity exhibited by water molecules, which attracts them to regions of very small, transitory charge on the oxygen molecules. This attraction is sufficient to hold oxygen in the water as dissolved oxygen. Even a very small rise in temperature will disrupt these tenuous attractive forces and drive oxygen from the water. Consequently oxygen levels tend to be lower in warm water than in cold, and dissolved oxygen concentrations are said to be temperature dependent. Extensive fish kills often occur in the summer as a result of warm water temperatures, which lead to drastic reductions in dissolved oxygen levels.

The ability of water to dissolve ionic, polar, and nonpolar compounds is responsible for the most well-known characteristic of seawater: its high concentration of dissolved materials. This is a direct result of the polar nature of water.

Seawater

Seawater contains traces of all naturally occurring elements. The vast majority of these elements are derived from terrestrial sources, dissolve in rainwater, and are carried to the sea by streams and rivers.

When atmospheric water vapor cools, condenses, and falls through the atmosphere, it encounters and dissolves gaseous carbon dioxide, causing the rainwater to become slightly acidic. This acidity increases the ability of the water to dissolve many different substances. Consequently, when this water falls on terrestrial rocks and sediments, many of the soluble materials dissociate into their component ions. The dissociation is due to the formation of hydrogen bonds, which form between the water and the polar and ionic molecules composing the rock.

These dissolved materials, primarily calcium, sodium, and potassium are carried in their dissolved form into streams and rivers, which flow into the coastal seas. Most of the materials dissolved in seawater originate from terrestrial sources in this manner.

The only materials found in the sea not having a strictly terrestrial origin are the negatively charged chloride, bromide, and iodide ions. These ions have their origin in the deep ocean, where they enter the sea dissolved in water from the earth's interior. This water, termed **juvenile water**, enters directly into oceanic water in rift areas, (chapter 2). In these areas juvenile water from deep beneath the earth's surface, accompanied by molten rocks, comes into contact with the overlying seawater. The dissolved materials in the juvenile water—the chloride, bromide, and iodide ions—remain in their dissolved form when they enter the seawater.

Analysis indicates that the ions dissolved in the juvenile water are those that are absent from river water as the result of the weathering of terrestrial rocks and sediments. It has also been shown that the chloride ion concentration in juvenile water is virtually identical to the chloride ion concentration in seawater. This implies that the chloride ion concentration in seawater is likely due to the influx of juvenile water into the deep ocean. The water from terrestrial sources and the juvenile water are slowly, but thoroughly, mixed by the currents of the sea (chapters 10, 11). The materials dissolved from terrestrial sources, when mixed with the materials dissolved in the juvenile water, give seawater its characteristic salinity.

Table 7.1 summarizes the major constituents of seawater.

Salinity

The total amount of dissolved material contained in 1 kilogram of seawater is generally considered to be the **salinity** of that particular sample of water. Salinity is more precisely defined as the total amount of dissolved material that is present in 1 kilogram of water, assuming that all the carbonates have been converted to oxides and the bromides and iodides replaced by chloride, with the organic substances oxidized. Salinity is often expressed in parts per thousand and given the symbol %oo.

The materials dissolved in seawater are slowly, but completely, mixed by the currents with the result that the relative amounts of the major elements dissolved in seawater are essentially constant, regardless of the ocean in which these elements are found. This constant relationship of the major elements is known as the law of constancy of composition.

It can be seen that the chloride and sodium ions comprise over 85 percent of the dissolved materials in seawater. The taste of salt water is due to the presence of both of these ions. Rivers, although they contain a certain amount of dissolved materials, do not taste salty, since the chloride is essentially absent.

Table 7.1	Ion	%
The Major Constituents of Seawater	Cl^-	55.07
	Na^+	30.62
	SO_4^2	7.72
	Mg^{+2}	3.68
	K^+	1.10
	HCO_3^-	0.04

The law of constancy of composition holds true for offshore oceanic water. In coastal waters the percentage of these elements is not in a constant ratio. This is caused by a variety of factors, such as the input of terrestrially derived materials that are dissolved in river water and rainwater runoff. In addition the coastal ocean is frequently misused by man and often serves as a convenient disposal area for waste products. For example, in the New York–New Jersey metropolitan region, over 200 tons of human, commercial, and industrial solid waste are disposed of daily by ocean dumping. Much of this material ultimately dissolves in the water column, affecting the law of constancy of composition.

Salinity Relationships

Density has been defined as the relationship of mass to volume. Since salinity is a measure of the total amount of material that is dissolved in a given volume of water, there is a direct relation between salinity and density. In general, if a constant temperature is assumed, the higher the salinity, the greater the density of a water mass. Therefore if seawater is carefully added to the surface of a freshwater system, or any other system of lower salinity, the more saline, denser water will sink beneath the less saline water; conversely, any water of a lower salinity, when added to the bottom of a more saline system, will rise to the surface.

Salinity determines the positions that water masses will take. For example, consider a tank divided by a movable partition, with seawater on one side of the partition and fresh water on the other (fig. 7.6a). The seawater, with its higher salinity, would be denser and its molecules would be subjected to a greater pressure than those of the fresh water. Removing the partition would cause the seawater to

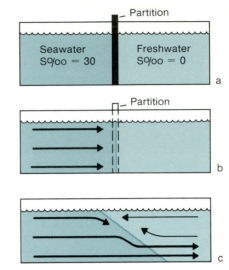

Figure 7.6 Denser water will move into regions of lower density (b). When it encounters less dense water, it will sink beneath it (c).

move from top to bottom into the fresh water in an attempt to reduce this pressure (fig. 7.6b). As the seawater moved, it would sink beneath the fresh water as a result of its greater density (fig. 7.6c).

The movement of water masses in response to salinity differences, which lead to density disparities, is responsible for the formation of many currents in the sea. These currents may be termed **pressure-gradient currents,** since they move in response to the pressure differentials between adjacent water masses.

Salinity Density Barriers

Differences in salinity, as well as the previously discussed temperature effects, cause water masses to layer, or stratify, in predictable patterns. The denser, more saline water will always be found at the bottom of a given water column. In the absence of temperature effects, there will be a vertical salinity distribution from the surface to the bottom of most marine systems, with water of intermediate salinity located between the very dense, highly saline bottom water and the less saline, least dense surface water.

Since the denser water sinks to and remains on the bottom, there is little mixing of the more saline with the less saline water. Indeed the difference in salinity is responsible for the formation of density

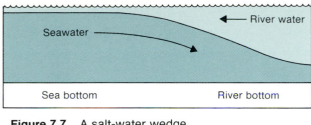

Figure 7.7   A salt-water wedge

The polar sea

barriers, which, in effect, physically separate the water masses. If the salinity differences are large, very strong barriers are formed and there is virtually no mixing of the water masses. Since these barriers are caused by differences in salinity, they can be termed **salinity density barriers** to distinguish them from the temperature density barriers.

Salinity density barriers markedly affect animal and plant distributions. In coastal areas where large rivers enter the sea, a salt-water wedge (fig. 7.7) may be formed as a result of saline marine water that encounters and sinks beneath the less dense river water that is flowing seaward. In these instances the surface waters are often completely fresh and support freshwater plant and animal populations. The deeper, more saline water, however, contains typical marine populations. In some cases this allows one to catch both fresh- and salt-water fish at the same location merely by changing the depth of the line.

Temperature Effects in Marine Systems

When seawater freezes, the dissolved materials are not incorporated into the ice crystals; rather, they remain in solution. Consequently as ice forms there is an ever-decreasing volume of liquid water, along with an ever-increasing amount of dissolved material, in the unfrozen water. Since density is the relationship of mass to volume, or $D = m/v$, and since in this case the salinity, and therefore the mass of the water, increases as the volume of the water decreases, there must be a corresponding increase in density. As ice formation increases, the salinity will continue to increase, causing the water to become denser. As a result of the relationship of ice formation to salinity and density, seawater, unlike fresh water, gets denser as it gets colder. The maximum

density of fresh water, on the other hand, is 4°C (39°F) because of the absence of significant quantities of dissolved materials.

There is a point, however, at which the increased concentration of dissolved materials makes additional ice formation impossible. This is due to the tendency of dissolved materials to inhibit the formation of ice. The vast majority of materials that are dissolved in seawater are present as either ions or polar molecules, and they form intermolecular attractions with the water molecules. This interferes with the ability of the water molecules to form hydrogen bonds among themselves and thereby enter into the crystalline ice structure. As a consequence additional heat must be removed from the system in order to bring the water molecules close enough to form hydrogen bonds among themselves.

The higher the salinity, the more dissolved material that is present within a water column and the lower the temperature must be before the ice can form. This phenomenon illustrates the freezing-point depression of water. This reduction in freezing point, combined with the insulation provided by the ice cover, prevents large quantities of seawater from freezing in the subzero temperatures of the Arctic and Antarctic.

As the surface of the sea absorbs heat from the sun, the molecular motion of the water increases, resulting in increased molecular motion and the subsequent disruption of hydrogen bonds. As additional heat is absorbed, some molecules attain sufficient energy to enter the vapor state. It is, however, only the water molecules that evaporate, while the material that is dissolved in the water remains in solution.

Due to the evaporation of only the water molecules, the dissolved materials will be present in an ever-decreasing volume of water. This is analogous to what occurs when sea ice is formed. The water volume decreases and, although the mass of the dissolved material remains essentially constant, the material is dissolved in a smaller volume of water, resulting in an increase in the density of the surface water. Eventually the water will reach a point at which it is denser than the water below, and it will sink to a depth of equal density. The increased temperature, however, may increase the volume of the surface water and thereby decrease its density to such an extent that this water, though highly saline, remains at the surface to overlie colder, less saline water. This occurrence is common in the tropics.

Since the dissolved material consists primarily of ions and polar molecules, intermolecular attractive forces develop between these dissolved materials and the liquid water. Hydrogen bonds, as well as these intermolecular attractive forces, must be broken before water molecules are able to enter the vapor state. The temperature at which large quantities of seawater evaporate or boil, is therefore higher than that needed to evaporate or boil fresh water. The increase in boiling point as a result of the presence of dissolved materials is called the boiling-point elevation.

Temperature and Salinity Relationships

The input of heat and rain to the sea does not occur uniformly in all of the world's ocean. In the equatorial and tropical seas, a high degree of solar heating does occur, leading to a high evaporation rate. In the equatorial seas and the tropical waters nearest the

Box 7.1 Hydrogen Bonds and Antifreeze

The tendency of materials to bring about both a boiling-point elevation and a freezing-point depression when dissolved in water is the principle that underlies antifreeze and coolants. These synthetic materials, when added to automobile radiators, prevent hydrogen bonds from developing and hinder ice formation. At high temperatures the intermolecular attractions between the solute and the solvent prevent the water from boiling, causing the solute to act as a coolant.

equator, the high annual rainfall counterbalances the water that is lost by evaporation. As a result highly saline water is not formed. In the more southerly tropics of the northern and southern hemispheres, respectively, as well as in the subtropical regions, the annual precipitation rate is much less. In these regions the high evaporation rate is not balanced by precipitation, and very saline water is formed. This water, however, generally remains at the surface, since it is less dense than the deeper, colder, water even though it may be more saline. Conversely, in the polar seas there is little heating, and a large quantity of water is found in the form of sea ice. These seas are also highly saline.

The markedly different temperature and salinity regimes in various regions of the world's oceans lead to significant density differences of these waters. It is the differences in density that both set water in motion and prevent water masses from mixing.

In the sea, temperature density barriers, as well as salinity density barriers form. In the majority of cases, the combination of highly saline and low-temperature water forms salinity/temperature density barriers. This is due to the tendency of salinity to increase with depth, while temperature decreases with depth. A mutual reinforcement in the density differentials is created, and these barriers become very efficient in preventing the vertical mixing of water

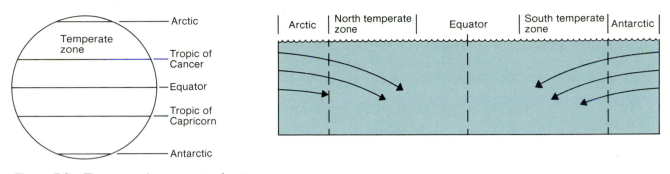

Figure 7.8 The general movement of water masses based on the different temperature regimes of the world's oceans

masses. As a result the denser bottom water is effectively prevented from mixing with the waters above.

Some vertical movement of water does occur, but it is generally a downward movement. This sinking water is also eventually prevented from mixing with the surface waters by density barriers.

Density barriers occur in the Sargasso Sea, which is located several hundred kilometers off of the east coast of the United States. In this area there is a high degree of solar heating, which leads to a high evaporation rate and a high surface salinity. Eventually the surface water becomes sufficiently dense and sinks beneath the colder, less saline water to a depth of equal density. This water then cools and becomes indistinguishable. As a consequence the subsurface waters become very dense and are separated from the upper water by strong density barriers.

A somewhat similar situation occurs when the Mediterranean Sea flows into the coastal Atlantic Ocean. The water in the Mediterranean Sea is subjected to relatively high atmospheric temperatures, causing a high evaporation rate that increases the salinity of this water. When the Mediterranean flows through the Straits of Gibraltar, it encounters the colder but less saline waters of the Atlantic Ocean. Since the Mediterranean waters are warm but highly saline, they are denser. They sink beneath the less dense oceanic water, forming a warm, highly saline subsurface water mass.

Low temperatures will also set water masses in motion. The low temperatures in the polar seas are responsible for the large quantities of sea ice that

form. Because of the exclusion of dissolved materials from this ice, the Arctic and Antarctic seas contain the coldest, most saline water in all of the world's oceans. These waters are therefore the densest in the world.

The more temperate seas that bound the Arctic and Antarctic waters are warmer (fig. 7.8). Since there is less ice formation, the waters are less dense. This natural system, composed of highly saline, very cold, and very dense water adjacent to warmer, less dense water is analogous to the artificial system described on page 116. Because of its greater density, the Arctic and Antarctic water will move from top to bottom into the less dense water mass. As the denser water encounters the less dense water, it will sink to the depth of equal density. Since these are the densest waters in the world, they will rapidly sink to the bottom. The Arctic water will move south, and the Antarctic water will move north as pressure-gradient currents.

These water masses are separated from the other waters in the ocean by remarkably stable temperature/salinity density barriers. Since these density barriers prevent mixing, the water in these masses tends to retain its characteristic temperature, salinity, and density for long periods and great distances. By measuring the temperature, salinity, and density of the waters, oceanographers have been able to trace the paths of these currents very accurately. The presence of density barriers and their tendency to effectively prevent the mixing of water masses has had profound effects on the distribution of life in the sea.

The secchi disk is commonly used to measure light penetration into a water column. The disk is attached to a line marked in meters and lowered into the water. The depth at which the secchi disk disappears from view is the extent of light penetration.

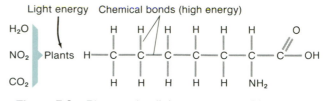

Figure 7.9 Plants using light energy are able to convert simple low-energy compounds into high-energy molecules. The light energy is converted to the more available high energy of chemical bonds.

Light, Density Barriers, and Life in the Sea

Life in the sea is dependent upon the plants that inhabit the sea. These plants, in turn, require sunlight, simple compounds called plant nutrients, and carbon dioxide, a gas dissolved in seawater. The majority of the plants of the deep ocean, termed phytoplankton, spend their entire lives suspended in the water.

The nutrients required by phytoplankton are the phosphate and nitrate ions, which are readily soluble in water. These nutrients and carbon dioxide contain very little energy, and so the phytoplankton use energy from the sun to convert these simple materials into high-energy chemical compounds. In the process, light energy is converted into chemical energy, which forms and is contained in the chemical bonds that hold these molecules together. In this case, the low-energy phosphate and nitrate ions and the carbon dioxide are converted into high-energy carbohydrates (fig. 7.9), and the light energy is captured and converted into more usable chemical energy.

When animals, commonly known as grazers, eat the plants, they obtain their energy by breaking chemical bonds and rearranging the carbohydrates that compose the plant tissue. A portion of this matter and energy becomes unavailable to the larger animals, since some of the plants and smaller animals will die before they are consumed and the smaller animals will also release wastes. The dead animals and plants and the waste products provide the energy source for the bacteria that break down these substances. As these materials undergo bacterial decomposition, plant nutrients are released into the water in their proper form to be reused by the phytoplankton.

The sun drives the entire process by providing the light energy. Sunlight is often termed **polychromatic light,** since visible light contains all the colors of the spectrum. Light is merely a form of energy, and when it enters water some of this energy is transferred to the water in the form of molecular motion. Since the penetration of light into the sea decreases with increasing depth, a depth will be reached where there will be insufficient light, and therefore energy, to allow plants to convert the simple plant nutrients and carbon dioxide into the more complex, high-energy forms. In other words there is a definite depth beneath which phytoplankton cannot exist.

Different colored light has different energies. As a consequence, light of a given color having a high energy will be able to travel deeper into the water than will a color of light of a lower energy. The blue-green portions of the spectrum have a higher energy than the red portions; therefore, blue-green light will travel deeper into the water than will red light. Many of the shrimp and fish that live below the zone of red light penetration are themselves red. Since red light is absent at these depths, it cannot be reflected; these animals therefore appear black and are camouflaged.

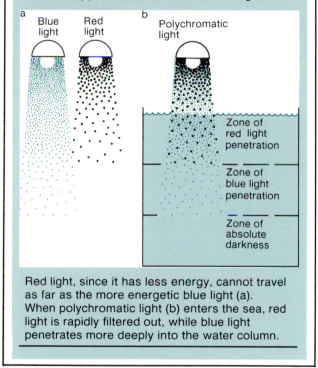

a Blue light Red light b Polychromatic light

Zone of red light penetration

Zone of blue light penetration

Zone of absolute darkness

Red light, since it has less energy, cannot travel as far as the more energetic blue light (a). When polychromatic light (b) enters the sea, red light is rapidly filtered out, while blue light penetrates more deeply into the water column.

The sea is often divided into two definite zones on the basis of sunlight penetration. The upper zone of effective light penetration is termed the **euphotic zone,** while the zone of darkness is often called the **dysphotic** or **aphotic zone.**

As noted, bacterial decomposition reconverts animal and plant materials back into the simple plant nutrients, phosphate and nitrate. In many cases, however, these plant nutrients are returned to the water beneath the euphotic zone—and often beneath a density barrier as well. Thus, although the plant nutrients may be in the proper chemical form, that is, **chemically available** for plant use, they cannot be used by the phytoplankton, since they are out of the euphotic zone and are therefore **spatially unavailable.** In other words the nutrients are in the proper chemical form to be used by plants, but they are in the wrong place—beneath the zone of effective light penetration. These spatially unavailable dissolved nutrients are frequently transported for long distances by deep-ocean currents (chapter 11).

Elements such as silicon are also required by many plants. These materials are not readily soluble in water and are therefore present in only trace amounts. The differential solubility of materials is due to the chemical characteristics of both the solute and the solvent.

The Implications of Differential Solubility

The ionic or covalent characteristics of the solute generally determines the degree to which it will dissolve in a solvent. As a general rule it can be said that like dissolves like. This means that an ionic or polar solute will be readily soluble in an ionic or polar solvent, whereas a nonpolar or slightly polar solute will be soluble in a nonpolar solvent, but only slightly soluble in a polar solvent such as water. Thus, sodium chloride is very soluble in water, while fats are only very slightly soluble. Fats are, however, very soluble in nonpolar solvents, and many nonpolar solutes are very soluble in fats.

Many man-made materials such as pesticides and plasticizers are essentially nonpolar and only slightly soluble in water. These same materials are very soluble in nonpolar solvents such as oils and fats. The tendency of a substance to dissolve more readily in one solvent than another is called differential solubility.

The biological magnification of DDT caused an alarming decline in bald eagle populations.

The differential solubility of solutes is responsible for the serious problems associated with the release of man-made, nonpolar compounds into the environment. For example, the pesticide DDT is nonpolar and only slightly soluble in water but is very soluble in nonpolar materials such as plant and animal tissues. As a result when DDT was applied, large amounts found their way into marine systems but only very small amounts dissolved in the water, while the remainder settled on or were incorporated into the bottom sediments.

The microscopic plants, in the process of removing and assimilating dissolved nutrients, indiscriminately took in the dissolved DDT. Since this material is more soluble in plant tissue than it is in water, it accumulated in the plants. As DDT concentrations in the water decreased, additional DDT from the bottom was able to dissolve in the water. This newly dissolved material also entered into and accumulated in the plants. The grazers feed on thousands of these plants in their lifetimes, and DDT is equally soluble in animal tissue. As a result DDT concentrations reached high levels in the grazer populations. The grazers, in turn, serve as food supply for various predators, which consume thousands of grazers in a lifetime. Thus, the DDT reached even higher levels in these populations. As a result of this pathway, DDT ultimately accumulated in large concentration in the animals.

In many coastal areas, birds such as eagles and ospreys are the final consumers, and they ultimately accumulated high concentrations of DDT in their tissues. This material interfered with the birds' ability to utilize calcium, a major component of egg shells. Subsequently many birds laid eggs with very thin or, in some cases, no shells, resulting in breaking of the eggs during the incubation period. In many areas of the United States, there has been a drastic decline in eagle and osprey populations, which has been directly related to the use of DDT. The tendency of materials to accumulate as they pass from plant to grazer to predator is known as **biological magnification** and is the direct result of the differential solubility of these substances.

Polychlorinated biphenyls, or PCBs, are toxic, nonpolar chlorinated hydrocarbons. These materials are widely used as heat exchange and insulating fluids in high-voltage electrical equipment. They are also added to paints, plastics, and rubber as stabilizers to inhibit bacterial growth and to make those materials resistant to decomposition.

PCBs have been implicated in birth defects in coastal bird populations. Studies by the American Museum of Natural History have revealed gross birth defects to be present in a tern nesting colony on Gull Island, New York, at the entrance to the Long Island Sound. Newly hatched chicks were found with four legs; others with deformed bills and legs. One chick was found with no left eye and a crossed bill. Analysis of both the eggs and the adult birds have shown that PCBs are present in very high concentrations.

These birds feed on small fish, mainly silversides and killies. The fish in turn feed on the primary consumer, zooplankton. It is probable that these substances were released into the environment from an undetermined source, entered the food chain, and accumulated in the top consumer, the tern.

Mercury follows a different pathway in marine systems. Mercury is present in the environment as a vapor and in its particulate form. It enters the environment by volcanic activity, the burning of fossil fuels, the use of mercuric pesticides, and the indiscriminate disposal of waste mercury in the sea. Mercury is virtually insoluble in both water and

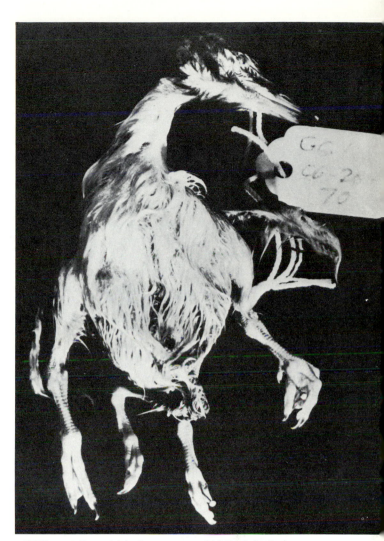

PCBs are thought to have caused birth defects in terns, typical birds of coastal areas.

biological tissue; as a consequence, it tends to sink to the bottom, since it is much denser than the water.

Once on the bottom, however, mercury is converted by microorganisms to a more soluble form, methyl mercury. The methyl mercury then dissolves in the water column and is absorbed on the cell walls of the phytoplankton. Grazing by zooplankton further accumulates this material, which then passes up the food chain. Also fish absorb methyl mercury through their gills and/or skin.

Methyl mercury causes deterioration of the central nervous system. Evidence also indicates that methyl mercury may cause genetic damage. There are several well-documented cases of methyl mercury poisoning from the eating of contaminated fish.

Summary

Because of its structure, the water molecule readily forms hydrogen bonds with four adjacent water molecules. Hydrogen bond formation increases as liquid water is cooled. This causes an increase in density with a decrease in temperature and leads to temperature stratification in both freshwater and marine systems. When combined with the stratification that results from the development of salinity differentials in marine waters, strong, stable, and persistent density barriers form, causing nutrients to become spatially unavailable for long periods of time.

The presence of temperature and salinity gradients in the sea also forms currents. These currents often transport nutrients that have become trapped beneath the density barriers, carrying them for long distances.

Its polarity makes water a remarkably versatile solvent. It is able to readily dissolve virtually all polar and ionic compounds, as well as many nonpolar materials. Many of these nonpolar materials are more soluble in solvents other than water, however. Hence, they will readily leave the water and dissolve in the more compatible solvent. This characteristic has led to the biological magnification of many deleterious compounds and has resulted in the chemical contamination of many organisms.

Review

1. Why would water tend to layer or stratify in a deep, temperate lake over a twelve-month period?
2. What is a temperature barrier? Why does it form?
3. Define a salinity barrier. How does it form?
4. What is meant by salinity?
5. Explain how temperature and salinity affect the density of water.
6. Why is it common to find higher salinity water at the sea's surface in the tropics?

References

Horne, R. A. (1969). *Marine Chemistry: The Structure of Water and the Chemistry of the Hydrosphere.* New York: Wiley Interscience.

Kuenen, P. H. (1963). *Realms of Water: Some Aspects of its Cycle in Nature.* New York: John Wiley.

For Further Reading

Black, J. A. (1977). *Water Pollution Technology.* Virginia: Reston Publishing Co.

Gross, M. G. (1977). *Oceanography: A View of the Earth.* New Jersey: Prentice Hall.

8
Biotic and Abiotic Relationships

Key Terms

abiotic factor
autotroph
benthos
biological oxygen demand
biotic factor
compensation depth
copepods
heterotroph
holoplankton
media
meroplankton

nekton
phytoplankton
phytoplankton bloom
plankton
primary consumer
seasonal succession
secondary consumer
sessile organism
substrate
tertiary consumer
zooplankton

The physical and chemical components of water interact with and affect the plant and animal life in the sea. Conversely the plants and animals, as a result of their life processes, modify the physical and chemical components of the water columns that they inhabit.

The differences in density between air and water are among the major factors responsible for the vast differences between terrestrial and marine organisms. The high density of seawater has caused marine organisms to evolve both structurally and functionally in far different directions than terrestrial forms.

Terrestrial environments are notable for their vast forests and grasslands in which both predator and prey move rapidly about. The seas of the world, on the other hand, have few large plants. None of these plants are equivalent to the redwoods of California, the prairies of Kansas, or the spruce forests of Maine; and few marine animals, either predator or prey, are capable of rapid movement.

Many of the differences between terrestrial and marine plants and animals are due, primarily, to the differences in density between air and water. As a result of the high density of seawater, the forests of the sea can consist of single-celled microscopic plants that form the foundations of life in the sea, and even the large marine plants, such as kelp, do not need rigid supporting tissues, since the water provides support and holds them upright.

Marine plants are capable of carrying out photosynthesis, which produces both oxygen and the food that all marine animals require. Various ions dissolved in the water are essential for the growth and reproduction of plants. The plants actively remove these materials from the water column and thereby alter its chemistry. They then modify the structure of these ions during photosynthesis and concentrate the materials within their tissues.

Since the majority of marine plants are suspended in the seawater, marine animals are surrounded by their food supply and have evolved lifestyles that are very different from their terrestrial counterparts. All animals are dependent, either directly or indirectly, on plants for a food supply and feed either on the plants or on other animals that feed on the plants. When the plants that have escaped the grazers die and the animals die and/or excrete wastes, these materials decompose and ultimately dissolve in the water. As a result of these interactions, the plant and animal life in the sea are constantly modifying the water by removing and adding material.

All plants require sunlight in order to carry out photosynthesis; however, seawater rapidly filters polychromatic light. This reduction in light drastically limits the extent of the euphotic zone; as a result, the plants are confined to the relatively small, upper sunlit portion of the sea. In addition the presence of plants, animals, and suspended sediments in the euphotic zone often physically blocks light and prevents it from penetrating into the water column. This further limits the amount of the sea that is able to support plant life.

The depth of effective sunlight penetration controls the availability and utilization of the dissolved ions that are required by the plants. This factor, in turn, regulates the life-styles of marine plants and animals.

Biotic and Abiotic Factors

It is convenient to divide both terrestrial and marine systems into their **biotic** and **abiotic** components, or **factors**. The biotic factors consist of the plants and animals that inhabit a given area, as well as the bacteria that decompose the dead animals, plants, and excretory material. Thus, the biotic component refers to, and is defined as, the living portion of a system.

The abiotic, or nonliving, components consist of the **media** and the **substrate.** The medium is defined as the material that immediately surrounds the organism, while the substrate is considered to be the material in or on which an organism lives. There are only two media—the air and the water. Therefore, on the basis of media, it is possible to divide the earth into two major areas: the terrestrial and the aquatic, including marine systems.

In marine systems the medium itself contains other abiotic factors that affect the properties of the water column: the dissolved ions required by the plants for use in photosynthesis and trace elements and contaminants, such as pesticides. Collectively these dissolved solids give seawater its characteristic salinity.

Salinity and temperature are abiotic factors that, either separately or in conjunction with each other, affect the density of water masses. Since density differences of adjacent water columns result in the formation of pressure-gradient currents (chapter 7), the density of the medium is an abiotic factor of the utmost importance. Light and dissolved gases, such as carbon dioxide and oxygen, are other significant abiotic factors in marine systems. Sunlight is, perhaps, the most important abiotic factor, since solar energy drives photosynthetic reactions and thereby converts simple, low-energy materials into complex high-energy carbohydrates. Energy from the sun also increases the molecular motion of the waters of the sea, which in turn increases the temperature of the seawater.

Temperature has major effects on air–sea interactions. As noted in chapter 7, air warms at the surface of the sea, becomes less dense, and rises into the upper atmosphere, where it cools. The cooler, denser air then descends and ultimately forms the trade winds and westerlies, which generate the major wind-drift currents of the world's seas (chapter 10).

Water has a significantly higher density than air; as a result, many marine organisms, unlike any terrestrial organism, can live suspended in their medium. It is the density difference between air and water that is responsible for many of the remarkable differences between terrestrial and marine organisms.

The substrate is variable in both terrestrial and marine systems. In terrestrial areas the substrate may range from moist woodland soils to the sandy substrate of beaches and dunes. The sediments found in marine systems range from exposed bedrock and boulders to the finest clays and silts. The size of coastal sediments is generally dependent upon the degree of wave shock, the currents, and water movements in a given area.

The substrate of a particular area will profoundly influence the distribution of marine organisms. For example, in sandy areas where the substrate shifts slightly with each breaking wave, burrowing organisms such as clams are found; in silty areas of slow-moving water, mud snails predominate; while on rocky, wave-washed coasts, animals such as barnacles, which are capable of maintaining their positions by cementing themselves onto the rocks, are common. Thus, the type of organisms found in a given area are dependent upon the sediment type, which in turn is determined by other abiotic factors, such as wave shock and currents.

The interactions of the biotic and abiotic components within a system influence and change each other. The summation of these interactions results in the environment of a given area. Since these interactions are continuous, the environment is also constantly changing, generally in subtle ways. Thus, the animals and plants that compose these systems are continually reacting and adjusting to these changes.

The Biota of the Sea

The biotic components, or biota, of any marine system consist of the plants, animals, and bacteria that inhabit a given area. The plants are often termed the producers or **autotrophs,** since they are capable of converting carbon dioxide and the simple phosphate and nitrate ions into complex, high-energy carbohydrates in the process of photosynthesis. Oxygen is released as a by-product during this process.

The animals, often referred to as the consumers or **heterotrophs,** are not capable of synthesizing carbohydrate from simple, low-energy materials, as are the plants. They are consequently totally dependent on the plants as a food and energy source. In addition the animals use the oxygen produced by the plants in the process of respiration, during which these high-energy compounds are utilized. The grazers, which feed directly on the plants, are often called herbivores and are the **primary consumers.** Other animals obtain their food by feeding on the grazers. These animals are considered to be the carnivores or **secondary consumers.** Larger carnivores that feed on this group of consumers are termed **tertiary consumers,** and so forth.

The bacteria decompose the dead animals and plants, as well as the excretory material released by the animals. A portion of this material is used by the

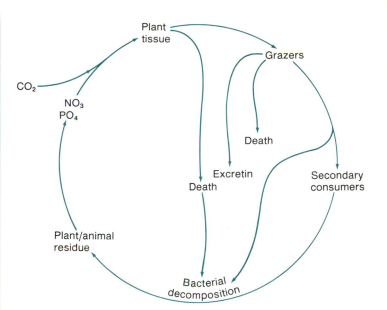

Figure 8.1 Simple, low-energy materials are taken in by plants, are converted to plant tissue, and serve as an energy source for various animals. When the plants die and the animals die or release waste products, these materials are utilized by the bacteria as an energy source. Phosphate and nitrate are thus returned to the water column.

Jellyfish spend their lives suspended in the water column.

bacteria for their own life processes, while the remainder is returned to the water column as nitrate, phosphate, and various other materials (fig. 8.1).

Two major forms of plant life are found in the sea: those attached to or rooted in the substrate and those that float on or are suspended within the water column. The attached forms are obviously capable of maintaining a fixed position, while the floating and suspended forms are moved passively about by the winds and currents. The distribution of these forms is controlled by the availability of sunlight.

In the deep ocean, light penetrates only into the upper waters; as a result, only the floating or suspended forms are found in these areas. In shallow coastal seas, light often penetrates to the bottom, and in these areas the rooted and attached forms are present along with the floating and suspended plants (fig. 8.2).

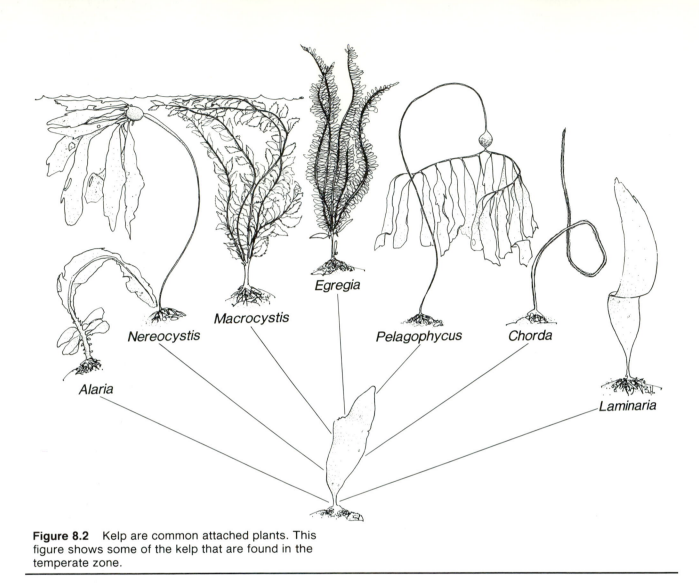

Figure 8.2 Kelp are common attached plants. This figure shows some of the kelp that are found in the temperate zone.

In the open ocean the vast majority of the plants are microscopic and are suspended within the water column. Since these plants are microscopic, most of the grazers are also small. Both the single-celled plants and small grazers are considered to be a part of the **plankton** of the sea. The term plankton refers to any organism, regardless of size, that is free-floating or drifting and that has its movements controlled by the motion of the water.

The larger animals, generally the carnivores, that are able to control their position in the water column are termed the **nekton.** Common examples of nektonic animals are herring and sharks. Plants and animals that live on or in the bottom constitute the **benthos.** As noted before, benthic plants are relatively uncommon in the sea, since only a small portion of the bottom is within the euphotic zone. These plants are confined to the shallow coastal seas, where they can find a suitable substrate for attachment and sufficient light to permit photosynthesis. The red algae called Irish moss, the brown kelps, and many green seaweeds are common examples of coastal benthic plants in the temperate zone. Since benthic animals are not dependent upon sunlight, they can be found in both the deepest oceans and shallowest seas. For example, the glass sponge is found at depths of over 2,500 meters (8,200 feet), while the common sea star of the west coast is found in the shallow coastal waters.

Hammerhead sharks are widely distributed in all of the warm seas of the world ocean. They are commonly found around islands, on the continental shelf, and further offshore.

The limpet *Acmaea sp.* is a common inhabitant of the shallow waters of the west coast. They attach so firmly to the substrate that an estimated force of 70 pounds per inch is needed to remove them.

In shallow coastal areas, where sunlight penetrates to the bottom and the substrate consists of rocks or pebbles, a film of algae generally coats these sediments. Benthic animals, such as the common periwinkle snail on the east coast and the limpet on the west coast, move over the substrate and scrape off the algae with a filelike mouth called a radula. In areas where the substrate is silty and in or on substrates beneath the euphotic zone, burrowing or surface-crawling animals ingest the substrate, digest the usable portions, which are generally the fragments of dead animals and plants, and release the remainder with their excretory material. These animals are the deposit feeders and include the common sea cucumbers of the east coast.

Because of the great depth of the oceans, much of the bottom is far below the euphotic zone, and as a result the vast majority of plants are the single-celled plankton that are confined to the upper sunlit portions of the water column. These plants are suspended in the medium and serve as a food supply for the small animal plankton. Since these animals live in a medium that not only contains but also surrounds them with their source of food, they have developed structures, termed setae, that enable them to screen or filter the microscopic plants from the water. The planktonic grazers, in turn, serve as a food supply for several secondary consumers. These carnivores have also developed structures that enable them to filter the grazers from the water in which they are suspended. Collectively the grazers and the secondary consumers are called filter feeders.

It is apparent that the seas of the world contain several forms of animal that have developed specialized methods of obtaining food. The deposit feeders extract their food from animal and plant residues that are incorporated into the silts that they ingest. The snails and limpets use filelike structures to scrape algae from the coarser sediments. And the filter feeders strain the food from the media.

As noted previously, the vast majority of the plants in the sea are the single-celled plankton. These plants provide the food and energy for the animal plankton. The animal plankton, in turn, are fed upon by a wide variety of animals, whose remains, along with those of the secondary consumers, provide the

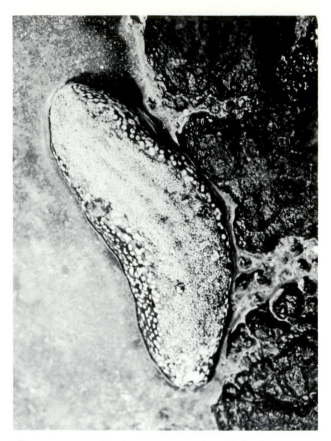

The sea cucumber is a common deposit feeder.

majority of food for the deposit feeders, particularly those in the deep ocean. Consequently both the plant and animal plankton are extremely important to the food relationships in the world's seas.

The Plankton

Plankton comprise a wide variety of plants and animals that are incapable of directed, sustained movement and hence are unable to effectively orient themselves within the water column. Plankton can range in size from the large, macroscopic seaweeds and various jellyfishes to the microscopic, single-celled plants and animals. The most important plankton, in terms of food supply, are the single-celled plants and the small, multicellular animals that feed upon these plants.

These plankton are commonly divided into plant plankton, known as **phytoplankton,** and animal plankton, called **zooplankton.** The microscopic

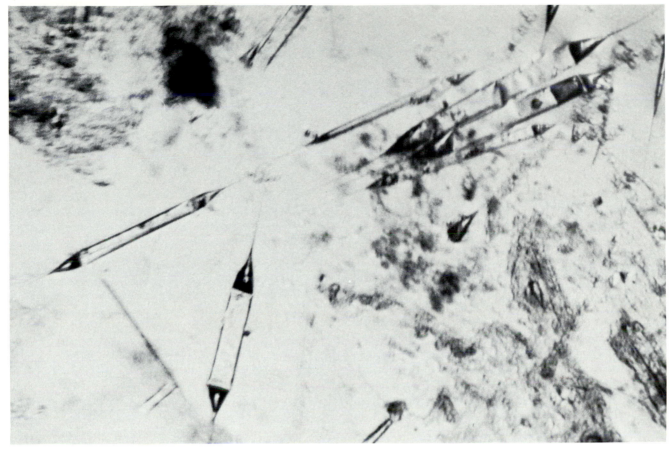

Phytoplankton

phytoplankton and small multicellular zooplankton are commonly collected for study by using nets of various mesh sizes. Consequently they are often divided into five categories on the basis of size only, irrespective of their being phytoplankton or zooplankton. The megaplankton are those organisms that are greater than 2 millimeters (0.08 inch) in size. Macroplankton range from 2 millimeters to 0.2 millimeter; while microplankton are between 0.2 millimeter and 20 micrometers in size (a micrometer is one-millionth of a meter—approximately .000039 inch). The smaller nonoplankton range between 20 micrometers and 2 micrometers, while the ultraplankton are the smallest and are less than 2 micrometers in size. Recently, a sixth category of plankton was discovered—the picoplankton. Picoplankton are less than 0.2 micrometers in size. Subsequent analysis has shown that these organisms contain chlorophyll; thus, they are phytoplankton.

Some organisms, known as **holoplankton**, spend their entire lives as plankton; the single-celled phytoplankton, for example, are considered to be holoplankton. Other organisms, the **meroplankton**, spend only a portion of their lives as plankton. Meroplankton are generally the young, larval stages of various animals. Upon reaching their adult stages, these animals spend their lives on the bottom or swim about in the water column. Common examples of meroplankton are the larval stages of fish and oysters. Not all zooplankton are meroplankton, however; organisms such as the jellyfish spend their entire lives as holoplankton. In addition a large group of crustaceans, the **copepods,** are holoplankton. The copepods are extremely important, since they are the major consumers of phytoplankton and, in turn, serve as a food supply for larger organisms.

Phytoplankton

Phytoplankton consist of single-celled plants that are between 1 millimeter and 0.1 micrometers in size. They include the diatoms, the coccolithophores, and the chrysomonads, all of which comprise the chrysophytes, the dinoflagellates, the silicoflagellates, the cryptomonads, the green algae, and the blue-green algae.

The diatoms and dinoflagellates are the phytoplankton that are commonly collected during periods of rapid phytoplankton growth and reproduction. Such periods are termed **phytoplankton blooms.** The diatoms actively remove silica from the water column and incorporate this material into an external covering called a frustule. Since diatoms are able to divide up to three times per day under optimal conditions, they are capable of removing large quantities of silica, as well as plant nutrients, from the water during phytoplankton blooms.

In addition to the importance of the dinoflagellates and the diatoms as a food supply, certain species of dinoflagellates are responsible for causing the deadly red tides. Dinoflagellates from the genera *Gonyaulax* and *Gymnodinium* produce potent toxins, which are released into the water column. These toxins, as well as the organisms themselves, are taken in and concentrated by shellfish. Human consumption of the contaminated shellfish often leads to paralysis and/or death. If the populations of these dinoflagellates reach sufficiently high numbers during phytoplankton blooms, the toxins released may lead to a high mortality in fin and shellfish populations.

Less common phytoplankton are the blue-green algae, the coccolithophores, and the silicoflagellates. The blue-green algae are common in tropical waters, where they convert atmospheric nitrogen gas into the plant nutrient, nitrate. The coccolithophores remove dissolved calcium from seawater and use this element to produce an outer protective covering of calcium carbonate discs. The silicoflagellates, like the diatoms, remove silica from the water to produce a protective covering.

Box 8.1 Phytoplankton Flowers

Phytoplankton blooms recall to mind a time when a species of phytoplankton underwent a larger than normal bloom. This type of phytoplankton released a metabolic by-product that was a potent toxin, harmful to marine life and humans. A radio newsman, in discussing the problem, stated that the toxin was released when the plankton flowers opened! The term plankton bloom is, perhaps, unfortunate; the plankton do not produce blossoms during this time. Rather, the term refers to the rapid reproductive rate of phytoplankton when the euphotic zone is virtually filled with these organisms.

Various forms of phytoplankton remove certain materials from the water column during their periods of growth and reproduction. As a result, and depending upon the type of phytoplankton present during a given period, certain predictable materials will be removed from the water column and others, such as the toxins produced and released by the dinoflagellates, may be added. Thus, the phytoplankton influence the chemistry of a given water column, as well as provide a source of food for the animals, primarily the zooplanktonic copepods.

Zooplankton

The zooplankton are a diverse group of organisms that consist of the transitory meroplankton and the permanent holoplankton. From the aspect of marine food relationships and the transfer of energy to the larger animals, the holozooplanktonic copepods are of the utmost importance. These animals are actually crustaceans that vary in length from one to several millimeters. The copepods are the major grazers on the phytoplankton and, as a result, are involved in virtually every aspect of the sea's food relationships.

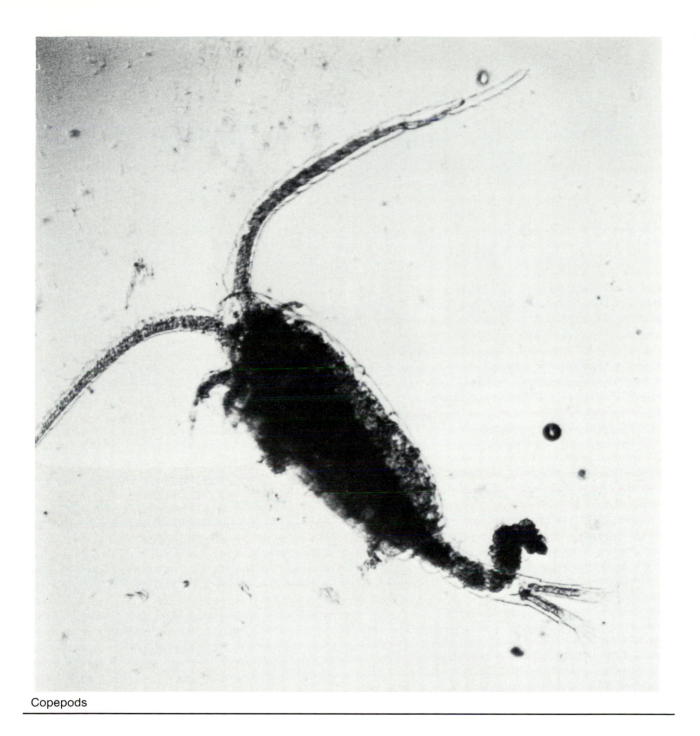

Copepods

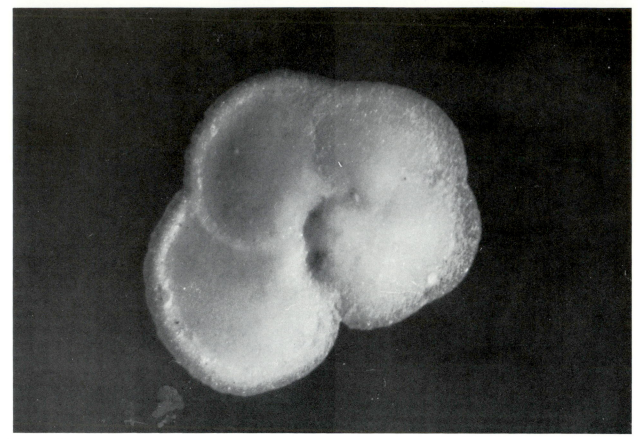

A foraminifera test

The population peaks of copepods closely follow the phytoplankton blooms. In fact it appears that it is the increase in copepod populations, rather than the depletion of the plant nutrients, that limits the extent of phytoplankton blooms. This observation is supported by occurrences in the polar seas, where nutrient levels remain consistently high throughout the year. In these areas the decline in the diatom populations in June occurs shortly after the copepods begin their population increase. Moreover copepods return the plant nutrients, phosphate, and nitrate to the water as excretory by-products. Since this material is returned to the euphotic zone, the continual replacement of these nutrients should serve to sustain the duration of the phytoplankton bloom. However, this is not the case since, even with this continual return, the number of phytoplankton decrease drastically as grazing by the copepods intensifies. The copepods, in turn, serve as a food supply for many of the larger marine animals and thereby serve to transfer the sun's energy, transformed by the phytoplankton, to these organisms.

Although not important from the aspect of energy transfers, the zooplanktonic and benthic foraminifera are important to a wide variety of oceanographic studies. The foraminifera are a form of single-celled marine animals that secrete a protective shell, or test, of calcium carbonate about their protoplasm. When these organisms die, the protoplasm decomposes, but the test remains and becomes incorporated into the bottom sediments.

Analysis of sediment samples (chapter 3) will reveal the presence of the tests. Since each species of foraminifera produces a distinctively shaped test, it is possible to determine which species are present in the sample. More importantly, many species of foraminifera are sensitive to changes in the abiotic environment. As a result some species proliferate, while others die out under different conditions.

The tests of foraminifera are actually microscopic fossils, termed microfossils, that provide a record of past conditions in the sea. For example, the foraminifera *Rotaliella heterocaryotica* has an optimal growth rate in water whose salinity ranges between 26 and 30 $^o/oo$. If the tests of this species are found in old sediments, it is possible to infer the salinity of the sea when that sediment was deposited.

The Autotrophs and Light

As noted in chapter 7, sunlight is considered to be polychromatic light, since it contains all of the colors of the electromagnetic spectrum. When sunlight enters the sea, less energetic red portions of the spectrum are rapidly filtered out and only the highly energetic blue-green portions penetrate deeply into the water column. Since the red light is preferentially absorbed by green plants, such plants must be confined to the zone of effective red light penetration in the sea in order to obtain the light necessary for photosynthesis. As a result green plants can occupy only the very upper portion of any water column.

Since red light is not abundant in the sea, various marine plants have compensated for its rapid loss. In relatively shallow coastal areas, three forms of large, attached, multicellular algae are present: the green algae, or chlorophytes; the brown algae, or phaeophytes; and the red algae, termed rhodophytes. The phaeophytes and rhodophytes are named for their pigments, brown and red, which mask the green chlorophyll. These brown and red pigments enable the brown and red algae to utilize various segments of the blue portion of the spectrum. Since this portion of the spectrum penetrates most deeply into the water column, these forms are able to live and photosynthesize at greater depths than are the green algae.

As a consequence the subtidal zone of almost any rocky coastline frequently exhibits three distinct zones of algae: an upper green zone, a middle brown zone, and a lower red zone. These zones correspond to the depth of penetration of the red and blue-green light into the water column.

In the deeper ocean the bottom is far beneath the euphotic zone and rooted forms cannot exist. In these areas the dominant plants are the phytoplankton. Phytoplankton, and indeed all plants, not only carry out photosynthesis, but also must use a portion of the energy produced during photosynthesis in respiration in order to reproduce, grow, repair tissues, etc. Generally, however, plants photosynthesize at a greater rate than they respire, and there is a net production of oxygen and food storage.

Phytoplankton must, obviously, remain in the euphotic zone in order to obtain the light that is required for photosynthesis. For each species there is a depth within the euphotic zone at which photosynthesis equals respiration. This is known as the **compensation depth** and is defined as the depth at which a given species of phytoplankton can survive but is unable to grow or reproduce. Below the compensation depth, respiration exceeds photosynthesis and there is a net use of the material that was produced and stored by the plant when it was above that depth. For this reason phytoplankton cannot survive for any extended period below the compensation depth, which varies with seasonal, daily, and environmental changes.

The angle at which the sun strikes the earth also varies seasonally. In the Northern Hemisphere the sun strikes the earth more directly in the summer and at a greater angle in the winter. As a result there is a greater penetration of light into the water column in the summer and the compensation depth is deeper. In the temperate zone of the Northern Hemisphere, the light-intensity in the winter is not sufficient to sustain a phytoplankton bloom. Thus, though the nutrients are generally sufficient in the winter, the phytoplankton bloom is delayed until the spring.

The angular relationship of the sun to the earth and its waters also varies daily. In the early morning and late afternoon, with the sun low in the sky, there is a greater amount of light reflected from the sea's surface than at noon. Thus, the compensation depth is greater at noon.

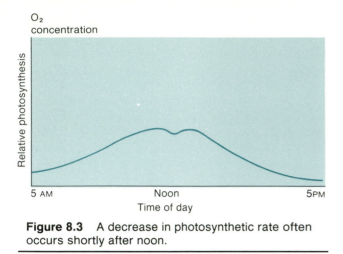

Figure 8.3 A decrease in photosynthetic rate often occurs shortly after noon.

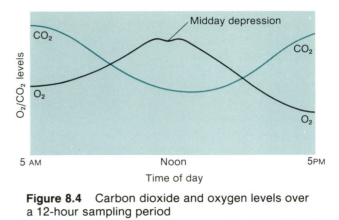

Figure 8.4 Carbon dioxide and oxygen levels over a 12-hour sampling period

There is not, however, a continual increase in photosynthesis with light-intensity. Rather, shortly after noon, particularly during the summer in the temperate zones, there is a midday depression in photosynthesis (fig. 8.3). This depression is in response to the maximum light that enters the water column at noon and is due to the destruction of photosynthetic enzymes by this excessive amount of light. It is, therefore, possible to reduce photosynthesis with too much, as well as too little, light.

During plankton blooms, the sheer numbers of organisms present in the water column will physically block the light. In these cases the plankton actually reduce the extent of the euphotic zone and raise the compensation depth. This tends to make less of the water column available for photosynthesis and may actually reduce the extent of the phytoplankton populations.

The amount of suspended sediments also affects the compensation depth. This generally occurs in coastal areas, where rivers bring large quantities of suspended sediments into marine waters. In addition sediments are carried into water bodies by runoff from terrestrial areas during and immediately after periods of heavy rainfall. These sediments physically block the light that enters a water column and reduce the compensation depth.

When the photosynthetic rate is reduced, less carbon dioxide is removed from the water. For example, if oxygen and carbon dioxide concentrations are graphed over a 12-hour period (fig. 8.4), it is found that the carbon dioxide levels are higher at night, when there is no photosynthesis occurring and both the plants and the animals are respiring. As the light intensity increases, the carbon dioxide decreases as it is removed from the water and used in photosynthesis. During this period, oxygen levels rise.

Thus, the amount of light that enters a given water column plays a role in the concentrations of dissolved gases within the water. In addition since water rapidly filters light, the vast majority of the sea is unsuitable for plant life and most marine plants are confined to the relatively small, sunlit surface waters. These plants must spend their lives either floating on the surface of the sea or suspended within the water column. The high density of seawater makes either life-style possible.

The Biota and Density

Since seawater is dense, small plants are able to live suspended within the water column. The presence of these small plants has allowed for the evolution of the small grazers, the zooplankton. It is important to realize that the zooplankton live suspended in a dense medium in which their food supply is also suspended. Thus, the primary consumers in the sea are surrounded by their food supply, and many have developed structures that enable them to filter this food from the medium. In terrestrial environments planktonic organisms would not be able to exist, since air,

Box 8.2 Phytoplankton and Cows: A Similar Principle

The dominance of phytoplankton in the open ocean may also be due to their rapid reproduction rate in comparison with the larger seaweeds. Some species of phytoplankton are able to reproduce up to three times in a single day. This is a distinct advantage in the open ocean, where currents can carry these organisms into inhospitable areas. Their rapid reproduction assures that some remain in the proper environment to continue the population, even though the majority are swept away.

The principle is similar to the bacteria that are found in the rumens of cows. These bacteria break down the cellulose eaten by the cow and convert it into fatty acid fragments. The fatty acid fragments are then passed into the cow's intestine, where they are absorbed to nourish the cow. Many bacteria are also passed "downstream" to the intestine, where they are destroyed. The bacteria, however, have such a high reproductive rate that a large population always remains in the rumen, even though the majority are killed when they are swept into the intestine. It is this population in the rumen that prevents the cow from starving.

about very slowly or, more commonly, they are non-moving or **sessile**. Marine systems are able to support large numbers of sessile or virtually sessile filter feeders. There are no terrestrial counterparts to the filter feeders.

Seawater also provides considerable support for marine organisms and holds them erect in the water column. As a result neither the single-celled plants nor the larger attached or floating forms have had to develop rigid, woody supporting tissues, and the animals that feed on these plants are able to consume virtually the entire plant. This is in contrast to terrestrial systems in which the plants have had to develop these supporting tissues to hold them erect. The presence of these tissues makes only a small portion of these terrestrial plants directly available to the animals as a food source, since the remainder is the unusable woody material.

Large surface-floating plants are rare in most of the world's oceans. The exceptions are the shallow coastal areas and the Sargasso Sea. These areas are notable for their calm conditions and the virtual absence of significant currents. In the open ocean, surface currents rapidly carry floating plants away from areas that are suitable for growth and reproduction; therefore, significant populations of these large surface-dwelling plants have been unable to develop in most of the waters of the world. Phytoplankton, on the other hand, are suspended within the water column and surface wind-drift currents do not tend to move these forms about as readily as they do the larger, floating plants.

because of its low density, would be unable to support even the smallest organism for even a fraction of its life cycle.

Since the phytoplankton and zooplankton are suspended in the water column, the larger animals are also surrounded by a medium that contains a food supply. As a direct result these organisms, like the zooplankton, have also developed structures that enable them to filter their food from the medium. Moreover many of the secondary consumers, particularly in the coastal areas, have not had to develop structures that would enable them to move about in search of food, and so these animals either move

The Biota and Water Chemistry

The life processes of marine plants and animals modify the water chemistry of the sea. During photosynthesis the plants remove dissolved carbon dioxide and various plant nutrients from the water. The major nutrients removed are the phosphate and nitrate ions. Sunlight provides the energy to convert these simple, low-energy materials into high-energy compounds, primarily carbohydrates. These are then used by the plants for growth, tissue repair, etc., while the excess is stored within the plants' tissues. Oxygen is produced as a by-product and is released into the water column.

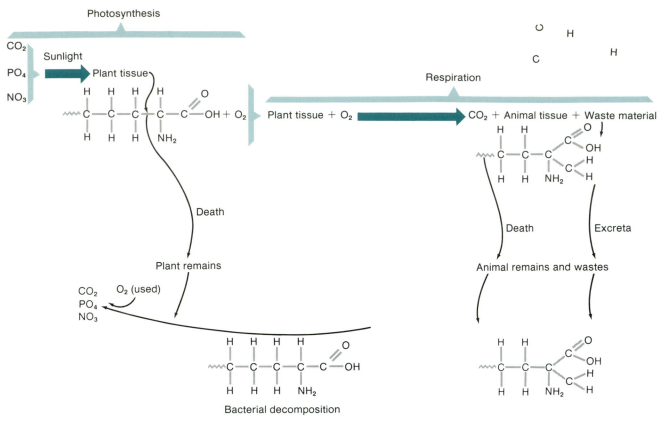

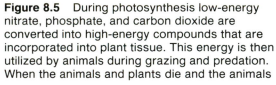

Bacterial decomposition

Figure 8.5 During photosynthesis low-energy nitrate, phosphate, and carbon dioxide are converted into high-energy compounds that are incorporated into plant tissue. This energy is then utilized by animals during grazing and predation. When the animals and plants die and the animals excrete wastes, these materials are used by the bacteria as an energy source, and the phosphorous and nitrogen that were bound up in animal and plant tissue are converted into nitrate and phosphate to be reused by the plants. In this example high-energy protein is formed, used by the animals, and decomposed by the bacteria.

Animals, unlike plants, cannot utilize nitrate, phosphate, and sunlight directly. Rather, the animals are totally dependent upon the plants to convert these materials into a form that they are capable of using: the fat, protein, and carbohydrate molecules. Animals therefore obtain their food and the energy necessary to carry out their life processes by feeding on the plants. The plant tissues, along with the oxygen produced in photosynthesis, are used by the animals in the process of respiration. During this process the animals convert the plant tissues into animal fats, proteins, and carbohydrates. A portion of these materials is stored by the animals; some is used for growth, reproduction, and the repair of tissues; and the remainder is released to the water column as carbon dioxide and waste products.

The carbon dioxide released by the animals is reused by the plants in photosynthesis, while the excretory material, along with the remains of dead animals and plants, sinks toward the bottom. Bacteria living within the water column or in or on the bottom sediments decompose this material and reconvert it into phosphate and nitrate ions (fig. 8.5).

Bacterial decomposition generally requires dissolved oxygen, which is removed from the surrounding water as the decomposition process progresses. An oxygen demand is placed on the system as a result of bacterial decomposition and the subsequent removal of dissolved oxygen from the water column. Since the material undergoing decomposition is biological in origin, this process creates a **biological oxygen demand,** or BOD. An increased BOD always implies the addition of ex-

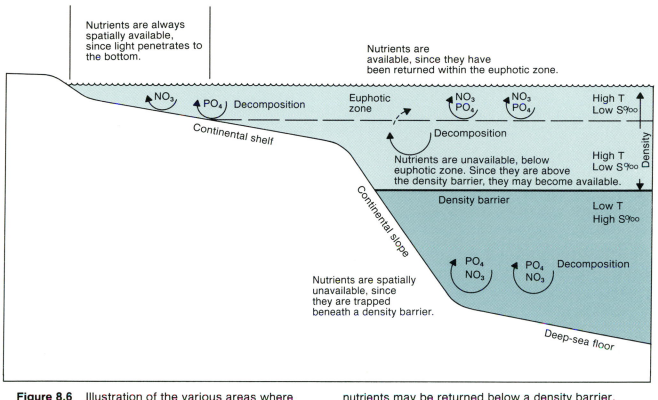

Nutrients are always spatially available, since light penetrates to the bottom.

Nutrients are available, since they have been returned within the euphotic zone.

NO₃   PO₄ Decomposition   Euphotic zone   NO₃ PO₄   NO₃ PO₄   High T Low S‰

Continental shelf

Decomposition

Nutrients are unavailable, below euphotic zone. Since they are above the density barrier, they may become available.   High T Low S‰   Density

Density barrier   Low T High S‰

Continental slope

PO₄ NO₃   PO₄ NO₃ Decomposition

Nutrients are spatially unavailable, since they are trapped beneath a density barrier.

Deep-sea floor

Figure 8.6 Illustration of the various areas where NO₃ and PO₄ may be returned to the marine environment by bacterial decomposition. Often the nutrients may be returned below a density barrier. Should this occur, these nutrients may remain spatially unavailable for long periods of time.

cretory material or plant and/or animal remains to a system. An increased BOD always results in the reduction of dissolved oxygen levels.

In addition to lowering the dissolved oxygen levels, bacterial decomposition breaks down plant and animal materials and converts them into the phosphate and nitrate ions that dissolve readily in the surrounding water. Thus, bacterial decomposition results in another alteration of the water chemistry due to the removal of oxygen and the addition of nitrate and phosphate to the water column.

The use of phosphate, nitrate, and carbon dioxide by the plants must occur in the euphotic zone, whereas the release of carbon dioxide and excretory material by the animals can occur anywhere in the water column, as well as in or on the bottom by the benthic animals. Similarly bacterial decomposition can occur in the euphotic zone, beneath it, and in or on the bottom sediments. Consequently bacterial decomposition may return plant nutrients to that portion of the water column that is below the compensation depth. Should this occur, the nutrients would be spatially unavailable and the plants would not be able to utilize these materials.

Water Density and Nutrient Availability

Bacterial decomposition begins as the plant and animal residues sink through the water column. Generally however, the majority of the phosphate and nitrate is returned to the subsurface waters, which are often denser than the surface waters and may be separated from the overlying water by a density barrier. If the subsurface waters are below the euphotic zone, the phosphate and nitrate, though in the proper chemical form for use by the plants, will be spatially unavailable to the plants. Additional decomposition will result in the return of more dissolved material to the dense subsurface water. This will tend to further increase the density of the water and strengthen the density barrier. Hence, the return of nutrients by bacterial decomposition influences not only the chemistry of the water but its density as well (fig. 8.6).

If the nutrients that are returned to the water column are spatially unavailable, they can be lost from the euphotic zone for long periods of time due to the stability and the persistence of the density barriers. In these instances they may be carried far from their point of input by subsurface currents. The loss of nutrients from the euphotic zone and their subsequent redistribution is quite common in the deep ocean (chapter 11).

In these situations, however, neither bacterial decomposition nor respiration generally results in the oxygen depletion of the dysphotic zone, since the sinking of oxygen-rich surface water carries sufficient dissolved oxygen to the deeper waters. Pressure-gradient currents play a major role in the redistribution of oxygen, as well as plant nutrients. When these currents encounter and sink beneath adjacent water masses of lower density, oxygen dissolved in the sinking water is carried beneath the euphotic zone and becomes available to the bacteria and animals of the dysphotic zone. Nitrate and phosphate are also carried out of the euphotic zone by this process and, as a result, these plant nutrients become spatially unavailable to the plants. In addition the evaporation of surface waters by solar warming increases the density of the remaining liquid water by increasing its salinity, since the dissolved materials remain in solution. As these waters become denser, they sink and carry dissolved oxygen, as well as nitrate and phosphate, to the dysphotic zone.

The life processes of marine plants, animals, and bacteria, in conjunction with the density differences of a given water mass, result in the redistribution of nutrients and oxygen in the seas of the world. Since temperature affects the density of water and since the density of water affects the distribution and availability of nutrients, phytoplankton blooms vary seasonally and geographically.

Phytoplankton Distribution and Nutrient Availability

The characteristics, extent, and length of plankton blooms vary geographically as a result of temperature, density, and salinity relationships. This is due to the tendency of water masses of different temperatures and/or salinities to stratify and form density barriers. These density barriers often trap nutrients beneath the euphotic zone and make them spatially unavailable to the phytoplankton. As a result there are marked differences in phytoplankton population dynamics in the coastal areas of the temperate, polar, and tropical seas.

Coastal waters in the temperate zone tend to stratify in the summer due to the formation of temperature and/or salinity barriers. These density barriers are generally beneath the euphotic zone; therefore, nutrients that are trapped beneath the euphotic zone in the summer are often spatially unavailable to the phytoplankton. The paucity of nutrients in the euphotic zone and grazing by the zooplankton generally result in low phytoplankton populations in the summer.

In the fall and winter, the water column cools in response to the lower air temperatures and eventually becomes homothermous. The uniform temperature of the water column, combined with the prevalence of storms that occur in the temperate zone during these seasons, disrupts density barriers and permits the efficient mixing of the surface and subsurface waters. Nutrients that were trapped beneath the summer density barriers are returned to the euphotic zone during the fall and winter months.

In the early fall this availability of nutrients leads to the fall phytoplankton bloom. As the numbers of phytoplankton increase, they remove the dissolved nutrients from the water column, incorporate them within their tissues, and thereby provide a food supply for the zooplankton, primarily the copepods. With the ready availability of phytoplankton, the zooplankton rapidly grow, reproduce, and serve as a food supply for the secondary consumers. Intense grazing by the copepods reduces the phytoplankton populations to their low winter numbers.

Nutrients are generally well distributed throughout the euphotic zone in the winter as a result of the mixing of the water column by winter storms. However, the low light intensity during this season prevents a winter phytoplankton bloom. In the spring a combination of high nutrient levels and increased light intensity leads to the spring phytoplankton bloom. As the phytoplankton populations increase, they again provide a food supply for the copepods, which, in response, undergo a zooplankton bloom. As in the fall increased grazing by the zooplankton serves to reduce the size of the phytoplankton populations (fig. 8.7).

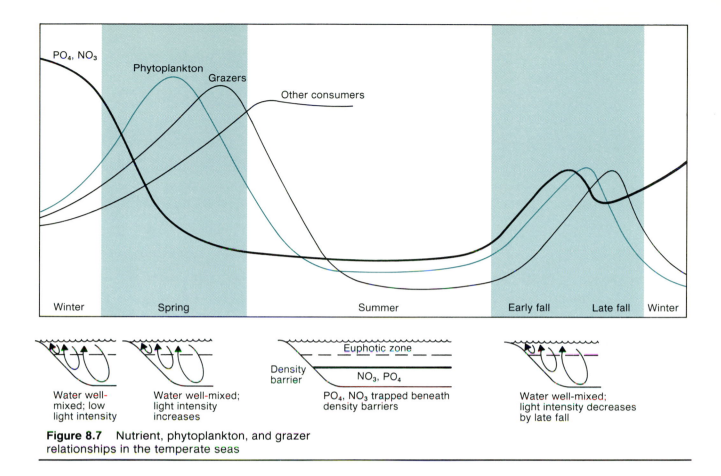

Figure 8.7 Nutrient, phytoplankton, and grazer relationships in the temperate seas

Since the excretory material released by the copepods consists primarily of nitrate and phosphate ions, which are the nutrients required by the phytoplankton, as the copepod populations increase, there is a continual return of these plant nutrients to the water column. During these periods the phytoplankton populations remain low because of the intense grazing by the zooplankton. As a result the nitrate and phosphate, though spatially available, are not used. Ultimately the copepods are either consumed by the carnivores or they die. In either case this source of nutrients to the euphotic zone is generally terminated by the late fall and the early summer, the times when the zooplankton blooms end in the temperate seas.

The excretory material and dead animals and plants that continually fall through the water column are decomposed by the bacteria that either live within the water column or in or on the bottom sediments. The water soluble nitrate and phosphate ions that are returned to the water column by bacterial decomposition are often trapped beneath the summer density barrier. These plant nutrients are therefore spatially unavailable for the phytoplankton until cooler temperatures and autumnal storms break down the density barriers and allow the nutrients to return to the euphotic zone so that the cycle may begin anew.

In polar areas stratification is seldom a factor in nutrient availability, since the entire water column is virtually homothermous throughout the year. Consequently density barriers seldom develop and, if they do, are generally of short duration. In general, phytoplankton populations in the polar seas are limited by light availability rather than by nutrient levels. For example, in the Arctic there is permanent night for four months during the winter, and there is a two-month period of continuous sunlight in the summer. As a result, a diatom bloom begins in May,

Density barriers are broken down by storms and hurricanes.

undergoes a slight reduction in June, and is followed by another bloom in July. There is only one zooplankton bloom, which begins in June and continues until the winter night begins in October. During this period nutrients are continually available in the euphotic zone.

In the tropics, the continually warm air temperature leads to a high evaporation rate of the surface waters (chapter 7). Although this increases the salinity of these waters, they do not sink, since the high temperatures cause a concomitant decrease in density, which keeps the saline waters at or near the surface. Thus, in the tropical seas the surface waters are warmer and less dense, though they often have a higher salinity. Permanent density barriers develop and effectively prevent the mixing of surface and subsurface waters in these instances.

Nitrogen and phosphorous are incorporated into plant and animal tissue and are carried out of the euphotic zone and beneath these density barriers when the plants and the animals die and release wastes. As a result nutrient levels in the euphotic zone of tropical seas are continually low. For example, nitrate

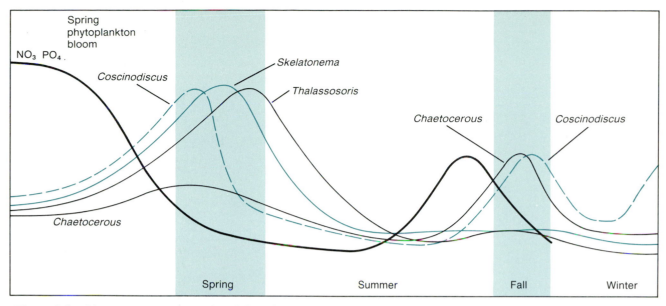

Figure 8.8 Phytoplankton succession. It is to be noted that the phytoplankton bloom actually consists of three species that become dominant at different periods in the spring and two different species that dominate in the fall.

concentrations in the euphotic zone are approximately 1/100 those of the least fertile waters in the temperate zone. Nitrate levels would be even lower, were it not for the presence of the blue-green algae in tropical waters. These algae convert atmospheric nitrogen gas to the more soluble nitrate ions. Phosphate concentrations are virtually undetectable in these waters. As a result, phytoplankton populations are continually low in tropical waters and consist primarily of microflagellates and coccolithophores.

It is only in the temperate zone that the density barriers are seasonal. This is in direct response to the seasonal atmospheric warming and cooling of these waters. Thus, only the temperate seas have two predictable, seasonal phytoplankton blooms.

In the deep ocean the density barriers are pronounced and so deep that winds and storms will not disrupt them. Consequently nutrients are permanently trapped beneath these density barriers and remain spatially unavailable, out of the euphotic zone. Deep-ocean currents traveling along the sea floor carry these dissolved nutrients for long distances and are responsible for their massive redistribution (chapter 11).

Phytoplankton Succession

In addition to the annual fluctuations in the total number of phytoplankton, there are also changes in the species that are dominant during any particular time. This seasonal change in the dominant species composition is predictable, repeated yearly, and is termed phytoplankton succession, or more appropriately, **seasonal succession.** As a consequence one or more species, generally diatoms and/or dinoflagellates, will be obvious dominants for a time, only to be replaced by one or more different dominant species.

In the Gulf of Mexico, diatoms such as *Thalassosoris* and *Chaetoceros* dominate in the spring; *Chaetoceros* and *Skelatonema* in the summer; and *Skelatonema* again in the fall. During the winter *Coscinodiscus* is the dominant phytoplankton. In the Long Island Sound, on the other hand, *Skelatonema* dominates in the spring, to be replaced by *Thalassosoris* in the late spring and early summer. *Chaetocerous* dominates in the late summer and early fall, to be replaced by *Coscinodiscus* in the winter and during the early portions of the spring bloom (fig. 8.8).

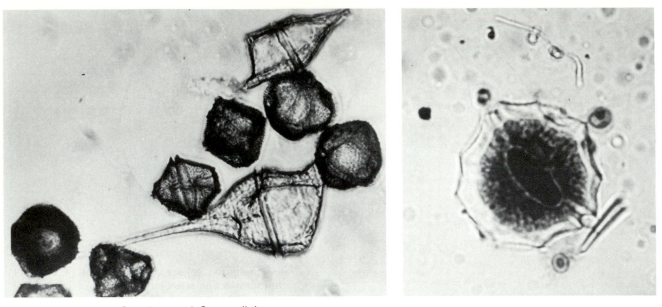

The dinoflagellates *Gonylaux* and *Gymnodinium* release toxins to cause the infamous red tides.

Several explanations have been postulated in an attempt to explain seasonal succession. Temperature has often been suggested as the major factor involved. However, certain dominant species reoccur at different water temperatures. Moreover temperature changes slowly in the sea, while the changes in dominant species are often rapid, with one species replacing another in a matter of hours at virtually the same temperature.

The utilization of nitrate and phosphate has also been suggested as a cause of seasonal succession. There is, however, little correlation between phytoplankton populations and nutrient levels, since the species change more rapidly than do nutrient levels. For example, in the Long Island Sound, nutrient levels are essentially the same when *Coscinodiscus* dominates as when that species is replaced by *Skelatonema*.

The active uptake of trace elements, such as silica by the diatoms and calcium by the *coccolithophores,* may be a factor in seasonal succession. This is most likely the case in the Sargasso Sea, where the *coccolithophores* and diatoms dominate in the spring. These organisms remove calcium and silica, respectively, from the water column and use these elements to produce protective coverings. These forms are replaced in the summer by the dinoflagellates, which do not require calcium and silica. The coccolithophores again dominate in the fall and winter in the Sargasso Sea.

A final possibility involves the release, rather than the use, of certain materials within the water column. It is to be recalled that the red tide is caused by the presence of the dinoflagellates *Gonylaux* and *Gymnodinium*. These phytoplankton release potent toxins into the water during their life cycles. Other types of phytoplankton may also secrete trace materials, termed metabolites, into the water column. These materials could then stimulate or inhibit the growth and/or reproduction of either that species or another species.

Summary

The plants, animals, and bacteria that inhabit a given marine area continually react with, change, and are changed by the total biotic and abiotic environment. Light plays a vital role in the sea, since the producers require sunlight as the energy source to convert the low-energy simple plant nutrients into the more complex high-energy molecules that can be used by the consumers. The dominant plants of the sea are the microscopic phytoplankton, while the dominant grazers are the copepods. The copepods are important, since they provide the major link in the transfer of energy from the phytoplankton to the large animals of the sea.

Density exerts profound effects on both the biotic and abiotic components of the sea. Many marine forms, both animal and plant, have developed unique life-styles in response to the high density of seawater. In addition density barriers are formed in the sea in response to temperature and/or salinity differences between water masses. In many cases these density barriers trap nutrients below the euphotic zone and make them spatially unavailable to the plants in these areas. This is the case in the tropical seas, as well as in all of the world's deep oceans, where the density barriers are permanent. As a result the phytoplankton populations are always low in these areas. In the temperate seas, the density barriers are seasonal, which results in seasonal plankton blooms. In the polar seas, density barriers do not tend to form, since the water column is virtually homothermous. In these areas phytoplankton blooms are controlled by the availability of sunlight.

Review

1. Compare and contrast biotic and abiotic factors. Give examples to illustrate.
2. Draw a typical food chain and label all components.
3. Differentiate between phytoplankton and zooplankton and between holoplankton and meroplankton.
4. Why are phytoplankton the major plants found in the sea?
5. Compare and contrast terrestrial and marine grazers.
6. Why is red light scarce in the sea?
7. What factors control photosynthesis in the sea?
8. How do the plants and animals modify the water chemistry of the sea?
9. How do nutrients become spatially available.
10. What are the major factors that control phytoplankton succession?

References

Colinvaux, P. A. (1973). *Introduction to Ecology.* New York: John Wiley and Son.

Hardy, A. C. (1954). *The Open Sea, Its Natural History: The World of Plankton.* London, England: Collins.

Odum, E. P. (1971). *Fundamentals of Ecology.* Pa.: Saunders.

Raymont, J. F. G. (1963). *Plankton and Productivity in the Oceans.* New York: Pergamon.

Russell-Hunter, W. D. (1970). *Aquatic Productivity.* New York: MacMillan.

For Further Reading

Colinvaux, P. (1978). *Why Big Fierce Animals Are Rare: An Ecologists Perspective,* New Jersey: Princeton U. Press.

Isaacs, John D. (1969, September). The nature of oceanic life. *Scientific American.*

9
Nutrients and Nutrient Cycling

Key Terms

biogeochemical cycle
buffer system
carbon cycle
detritus
downhill tendency of phosphorous
essential element
food chain
limiting factor
minor element

nitrate
nitrogen cycle
nitrogen:phosphorous imbalance
nitrogen:phosphorous ratio
phosphate
phosphorous cycle
trace element
trophic level

Plant nutrients tend to cycle throughout marine systems, from their simple, low-energy, dissolved forms to plants, animals, and then to bacteria. The plants assemble these simple materials into complex molecules that provide an energy source for the animals and bacteria, as well as for the plants themselves. Bacterial decomposition eventually reconverts these materials to their simple, low-energy forms and returns them to the water column.

In addition to the essential plant nutrients—phosphate, nitrate, and dissolved carbon dioxide—various forms of plants may require minor and trace elements. The absence of one or more of these required materials will limit the extent of phytoplankton blooms in the sea.

Nutrients may become spatially and/or chemically unavailable in marine systems. In the deep ocean and deeper areas of the coastal ocean, spatial unavailability is common and occurs when materials become trapped in the dense water below temperature and/or salinity barriers. In shallow bays the euphotic zone often extends to the bottom; hence, spatial unavailability of nutrients will not occur. Once nutrients become spatially unavailable in the deep ocean, few mechanisms are available to return them to the euphotic zone. As a result these materials may remain in the dysphotic zone for hundreds, if not thousands, of years.

Since the euphotic zone often extends to the bottom in shallow coastal areas, spatial unavailability of nutrients is rare. However, in shallow areas located in the temperate and polar regions, low light-intensity in the fall and winter generally limits phytoplankton populations. Thus, abiotic factors, other than the plant nutrients, often limit marine plant populations.

Only plants are capable of converting the simple low-energy nutrients into complex high-energy compounds by harnessing the sun's energy. Plants therefore are vital to and form the foundations of life in the sea. Food chains and food webs are often used to depict the relationships between the autotrophs and the various heterotrophic levels. In many cases, particularly in the deep ocean, relatively simple relationships exist, since the heterotrophs consist of only a few species. In coastal areas, particularly in salt-marsh and sea-grass communities, elaborate relationships exist, and the plants serve as a food/energy source for a wide variety of animals.

Marine plants, such as kelp, convert simple materials into a food source for the animals. The kelp appear as the dark areas beneath the water in this photo.

Biogeochemical Cycles

When considering the cycling of nutrients through any system, it is important to remember that matter and energy are interchangeable. Thus, in accordance with the first law of thermodynamics (chapter 6), matter and energy may be, and often are, converted from one form to the other. Consequently the terms nutrients, food, and energy are often used interchangeably.

All animal life in the sea is dependent upon the availability of plants, which serve as their food supply. Only the plants are able to utilize simple ions, molecules, atoms and sunlight and convert these materials, via photosynthesis, into the high-energy compounds that are required by both the plants and animals for their life processes. It is clear that the productivity of the sea is dependent upon a ready supply of these simple plant nutrients, as well as on an adequate supply of energy.

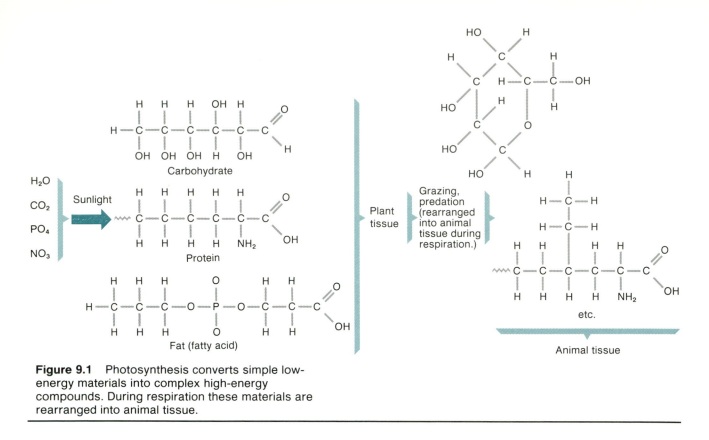

Figure 9.1 Photosynthesis converts simple low-energy materials into complex high-energy compounds. During respiration these materials are rearranged into animal tissue.

The plant nutrients tend to cycle from plant to animal to the bacteria. The process of bacterial decomposition then returns these nutrients to their chemically available form, to be used once again by the plants. Nutrients, therefore, are said to cycle from their simple, nonliving forms to the living components of a given system and then back into their nonliving forms. In other words nutrients cycle from the abiotic to the biotic and then back again to the abiotic components of a system. Cycles that encompass both the biotic and abiotic components of a system are termed **biogeochemical cycles.**

It is convenient to divide all of the nutrients into their constituent elements. On this basis the elements may be further subdivided into the **essential elements,** the **minor elements,** and the **trace elements.** The essential elements are those that are required by all animals and plants, regardless of the concentrations of those elements in the water. The

essential elements are carbon, oxygen, hydrogen, nitrogen, and phosphorous. These materials are generally available to the plants as liquid water, the dissolved gas carbon dioxide; and the dissolved ions **nitrate** and **phosphate.** In the process of photosynthesis and respiration, these materials are converted into carbohydrates, amino acids, proteins, and fats (fig. 9.1).

The minor elements are present in the sea in concentrations that range from one part per thousand up to one part per hundred. Elements such as magnesium, calcium, iron, and copper are minor elements. The trace elements are generally present in concentrations of less than one part per thousand. Manganese, silicon, zinc, and cobalt are important trace elements.

The minor and trace elements, unlike the essential elements, are not required by all marine organisms, but they are vital to the organisms in which they are found. For example, iron is found in the blood of many animals as a constituent of the hemoglobin molecule. However, the horseshoe crab of the temperate Atlantic and the Gulf of Mexico, requires

The horseshoe crab has blue blood due to the presence of copper rather than iron.

copper, rather than iron, and its blood is blue. Magnesium is essential to all plants, since it forms the central portion of the chlorophyll molecule. Silicon, on the other hand, is required only by some species of phytoplankton, such as the diatoms, and calcium is required by the coccolithophores.

The absence of one or more of the minor or trace elements will limit various species of phytoplankton. As noted in chapter 8, the depletion of calcium in the euphotic zone of the Sargasso Sea may limit the populations of coccolithophores in that area. The absence of one or more of the essential elements, on the other hand, will limit all of the phytoplankton populations. Obviously this would have severe impacts on the animal communities as well.

Limiting Factors

The size of any phytoplankton population is dependent upon a variety of factors. Light, the chemical and spatial availability of the essential elements, and

Box 9.1   Iron, Oysters, and Limiting Factors

Under natural conditions, nutrients are rarely spatially unavailable in coastal areas too shallow to form temperature–salinity barriers. However, alterations of the environment may result in severe nutrient depletions.

On eastern Long Island, New York, a shellfish farmer was having difficulty spawning oysters, since the salinity was too high. In an attempt to reduce the salinity, a deep hole was dug at one end of the shellfish spawning pond. The purpose was to intersect the water table and bring fresh water into the pond. The salinity did decrease, but the shellfish, rather than spawning, died.

Water analysis showed the phosphate concentrations to be very low, which led to a severe reduction in phytoplankton populations. The phytoplankton served as the food supply for the oysters and with virtually no phytoplankton in the ponds, the oysters soon starved.

Eventually it was determined that the acidic groundwater was rich in dissolved iron. When the groundwater entered the less acidic marine environment, the iron formed a precipitate and the precipitate actively removed phosphate from solution. The end result was a severe depletion in the oyster population. To rectify the problem the shellfish farmer found it necessary to periodically add a high-phosphate fertilizer to the pond.

---

magnesium are necessary for all plant life. The absence or low concentrations of these materials and/or the absence or low intensity of light will limit the size of the phytoplankton populations. The minor and/or trace elements may also be absent or in low concentrations. In these cases and with the exception of magnesium, which is required by all plants, generally only one or two species of phytoplankton will be affected. The paucity of these elements results, therefore, in phytoplankton succession (chapter 8) rather than in a reduction of the total phytoplankton in a given area.

The abiotic factors, sunlight, and the essential, minor, and trace elements serve to limit the productivity of the sea by imposing upper limits on phytoplankton populations. These interactions between the abiotic and the biotic components of the environment give rise to the concept of limiting factors. In any terrestrial or marine system having many interrelated components, the component that is present in the least amount is considered to be the factor that limits the biotic components; that is, the **limiting factor.**

In the temperate seas, light and nutrients are in short supply at different times; therefore, both serve to place limits on phytoplankton populations—but at different seasons. For example, in the winter the water is well mixed and the nutrient concentrations are high; yet the low light-intensities at this time prevent a phytoplankton bloom. In the winter, then, light would be the limiting factor. In the spring, light-intensity increases and light is no longer a limiting factor; the spring bloom occurs. In the summer the plant nutrients are incorporated into animal tissues, generally in the secondary or higher consumer stages. The nutrients are chemically unavailable to the phytoplankton at this season, even though light-intensity is high; the nutrients become the limiting factor in the summer.

In the polar seas of the Northern Hemisphere, there is perpetual darkness during the winter. At this season, although the nutrient levels are high and are spatially and chemically available, it is light that is the limiting factor. In the tropical and equatorial seas, however, the sun is at its most vertical position almost continually throughout the year and light is never the limiting factor. The low phytoplankton populations are attributed to the low concentrations of the nitrate ion and the virtual absence of phosphate (chapter 8). Since nitrate is present, phosphate is the limiting factor in these warm seas.

Box 9.2 Blue Blackfish

The tendency of fresh and/or low-density water to layer above coastal water and thereby make materials immediately available to the phytoplankton can adversely affect the food relationships in a given coastal area. In the Long Island Sound, which forms the boundary between Long Island, New York, and coastal Connecticut, large numbers of blackfish, secondary consumers, were found to contain such high concentrations of copper in their edible tissues that they were tinted a deep blue. The problem was traced to a local industry that was releasing small quantities of copper-enriched water into Long Island Sound. This material was inadvertently taken up by the phytoplankton, passed on to the grazers, and then to the blackfish. Shellfish in the area were also tinted blue by this type of biological magnification.

In one area there may be only one factor that continually limits phytoplankton populations, as is true in the polar and tropical regions. In other areas, particularly the temperate seas, the limiting factors may shift seasonally.

Nutrient Availability

Nutrients may be either spatially or chemically unavailable. When they are spatially unavailable, they are in the proper chemical form for use by the plants but are not in the euphotic zone. Spatial unavailability is common in the deep ocean, where these essential materials become trapped beneath density barriers. Since the density barriers are generally the result of both temperature and salinity differentials and are often far beneath the sea's surface, they are not easily disturbed by surface phenomena such as wind, waves, or storms. As a result the barriers are persistent, if not permanent.

Spatial unavailability is also quite common in the deeper coastal ocean. Generally, however, these areas are shallow enough and the density barriers near enough the surface to enable them to be broken down in the spring when the water column warms

and in the fall when it cools. Spring and fall storms facilitate the disruption of the density barriers during these seasons. Nutrients are returned to the euphotic zone seasonally in these areas.

In the very shallow coastal seas, sunlight is able to penetrate to the bottom; as a consequence, the larger attached plants, as well as phytoplankton, are present. In these areas dissolved nutrients are always spatially available and are frequently supplemented by nutrients that are brought in from terrestrial sources, such as river flow, surface runoff during rain storms, and so forth. The importance of the input of dissolved material from terrestrial sources is not to be underestimated. Since these nutrients are dissolved in fresh water, they are carried into the sea and will layer over the denser coastal waters at the surface of the water column. As a result these nutrients are readily available to the phytoplankton in the water column.

Spatial Unavailability and the Return of Nutrients

Nutrients become spatially unavailable in the deep ocean for long periods due to the stability of the density barriers. In the coastal ocean of the temperate zone, on the other hand, they become spatially available on a seasonal basis that coincides with the disruption of the density barriers in the fall and spring. Moreover nutrients are frequently brought into the coastal ocean from terrestrial sources.

Once nutrients do become spatially unavailable in the deep ocean, they are liable to be lost from the euphotic zone for hundreds, if not thousands, of years. Moreover there are very few mechanisms whereby nutrients are returned to inland areas where many, particularly phosphate, have originated. Nutrients may be returned to these areas by three mechanisms: upwelling, guano deposits, and by some types of spawning fish.

Upwelling occurs when surface coastal waters are moved offshore and replaced by the shoreward movement of deep-ocean water from the dysphotic zone. Because this deep water has been below the euphotic zone, the nutrients are generally high, since they had previously been spatially unavailable. Water movements of this type occur in areas where the

Guano deposits. Sea birds return large quantities of phosphorous and nitrogen to the terrestrial environment by excreting guano—the white material covering the rocks.

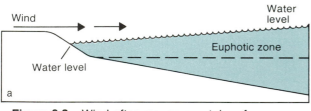

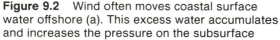

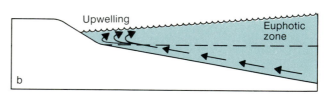

Figure 9.2 Wind often moves coastal surface water offshore (a). This excess water accumulates and increases the pressure on the subsurface water, which then moves onshore in response. Should this water encounter a rising shoreline (b), it may be brought back into the euphotic zone.

prevailing winds move the coastal waters offshore (fig. 9.2a). Eventually large amounts of coastal surface water are moved offshore and, being less dense, overlie the subsurface water. The additional weight of this water places added pressure on the subsurface water, which causes it to move onshore into an area of lower pressure (fig. 9.2b). Since this offshore water has been in the dysphotic zone, it has accumulated dissolved nutrients via bacterial decomposition, the sinking of nutrient-rich surface waters, and so forth. As the water moves onshore, it encounters a rising shoreline, and the nutrient-rich water is suddenly brought into the euphotic zone, where the nutrients become spatially available to the phytoplankton.

An upwelling of this type occurs at predictable times offshore Peru in the Pacific Ocean. The return of nutrients to the euphotic zone sustains a large phytoplankton bloom, and the plankton collectively serve as a food supply for the anchovies. These small fish, in turn, provide the major food supply for marine birds that roost on offshore islands. The birds

Pacific salmon migrating upstream. Man has inadvertently interfered with salmon migration by building dams. In many such rivers "ladders" have been constructed to enable the salmon to return to their spawning areas.

Salmon are considered to be anadromous since they return to fresh water to spawn.

excrete large quantities of a semisolid waste, called guano, which is rich in the essential elements, nitrogen and phosphorous. This excretory material dries and builds up to large quantities on these islands. Guano from these islands and similar coastal areas is often mined and used as a fertilizer.

Nutrients can be returned to inland areas by spawning fish. The Pacific salmon hatch in freshwater streams far from the sea and migrate to the deep ocean, where they grow and mature. As adults they return to the streams of their birth, where they reproduce and die. Since the salmon incorporate nutrients, grow and mature in the sea, and then return far upstream, nutrients are returned to inland environments. Fish that spend a portion of their life cycles in the sea and return to fresh water to spawn are termed anadromous.

Nutrients are returned fairly consistently to the euphotic zone in regions of upwelling and with much less consistency to the coastal and inland environments by guano deposition and by the spawning of anadromous fish. From the viewpoint of the total ocean and the quantities of nutrients that become spatially unavailable on a global basis, however, only a very small fraction are returned in comparison to the amount lost.

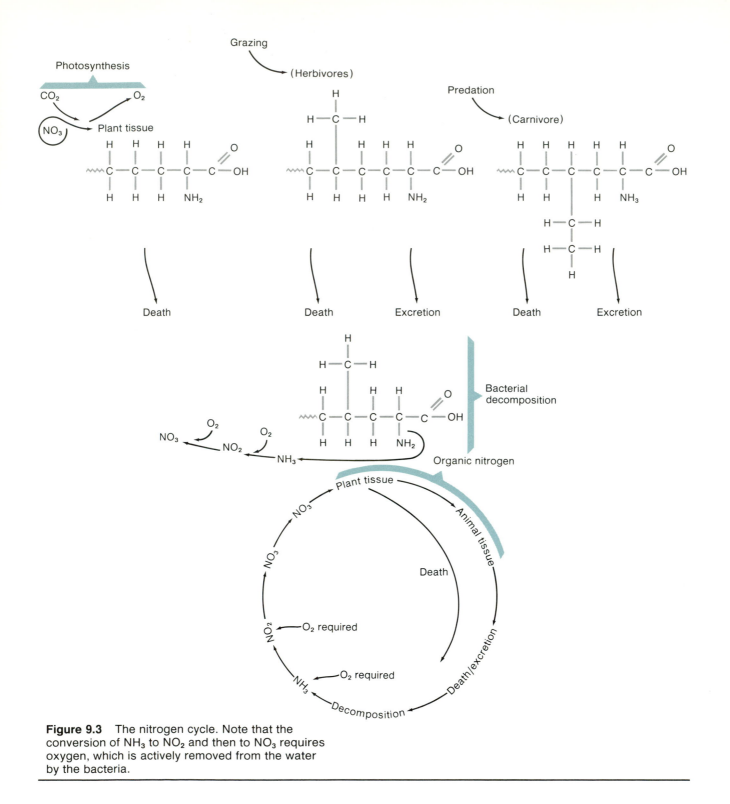

Figure 9.3 The nitrogen cycle. Note that the conversion of NH_3 to NO_2 and then to NO_3 requires oxygen, which is actively removed from the water by the bacteria.

The Nitrogen and Phosphorous Cycles

The essential elements are considered to be hydrogen, oxygen, carbon, nitrogen, and phosphorous, since they are found in all animal and plant tissues. These simple materials are taken in by the plants as water, carbon dioxide, nitrate, and phosphate and, in the process of photosynthesis (chapter 5), are converted into carbohydrates, fats, and proteins. Oxygen molecules are given off as a by-product.

Carbon dioxide provides the carbon and oxygen source for these complex molecules, and the hydrogen is supplied by the water molecules. The oxygen that is released as a by-product also originates from the water molecules taken in by the plants in the course of photosynthesis. Since water and carbon dioxide are readily available in the marine environment, the essential elements—carbon, hydrogen, and oxygen—are not limiting factors. Phosphorous and nitrogen, on the other hand, are frequently limiting.

Both nitrogen and phosphorous are found in marine systems in several different forms. In their simplest forms they are chemically available to the plants as the negatively charged phosphate and nitrate ions, termed inorganic nitrate and inorganic phosphate. In the process of photosynthesis, inorganic nitrate and phosphate become chemically bonded to long chains of carbon atoms and are thereby converted into what is termed organic nitrogen and organic phosphorous. Since animals are able to use the nitrogen and phosphorous in their organic forms, their grazing and predation transfers these materials from primary to secondary consumers. Upon the death of the plants and the excretion and death of the animals, the organic nitrogen and phosphorous are converted back into their inorganic forms.

The organic nitrogen undergoes a three-step bacterial conversion. Initially the nitrogen is removed from its carbon chain as water-soluble ammonia. The ammonia is then converted to the nitrite ion by nitrifying bacteria. In this process the hydrogen atoms are removed and replaced by two oxygen atoms. Since the source of this oxygen is the dissolved oxygen from the water column, this step places a biological oxygen demand (BOD) on the system. The third step involves the addition of another oxygen atom to form the nitrate ion. This process also places a BOD on the system. The organic nitrogen is converted to inorganic nitrate and in this form is again chemically available to the plants.

The organic phosphorous that is bacterially decomposed is converted in a single step into inorganic phosphate. As inorganic phosphate it is once again chemically available to the plants. The **nitrogen cycle** and the **phosphorous cycle** are summarized in figures 9.3 and 9.4.

The Sources and Fate of Phosphorous and Nitrogen

The majority of the phosphorous in the sea originates from the weathering of terrestrial materials. Phosphorous is present in terrestrial rocks and sediments as a constituent of ionic molecules, such as calcium phosphate. When polar water molecules come in contact with the calcium phosphate, it dissociates into positively charged calcium ions and negatively charged phosphate ions (fig. 9.5). Both of these ions are readily soluble in water. When dissolved in the water, they travel into rivers and streams and ultimately enter the coastal marine environment.

Since the water containing the phosphate is less dense than the saline coastal waters, it tends to remain at the surface in the euphotic zone. It is therefore both spatially and chemically available to the phytoplankton, is converted to organic phosphorous in the process of photosynthesis, and enters the biotic portion of the biogeochemical cycle. In shallow coastal areas the phosphorous cycle involves the transfer of phosphorous from plant to animal to decomposer. Phosphorous may cycle in this fashion several times in a coastal ocean.

Eventually, however, the phosphate may chemically combine with iron to form the insoluble iron phosphate molecule. When this occurs, the iron phosphate settles out of the water column and becomes incorporated into the underlying sediments,

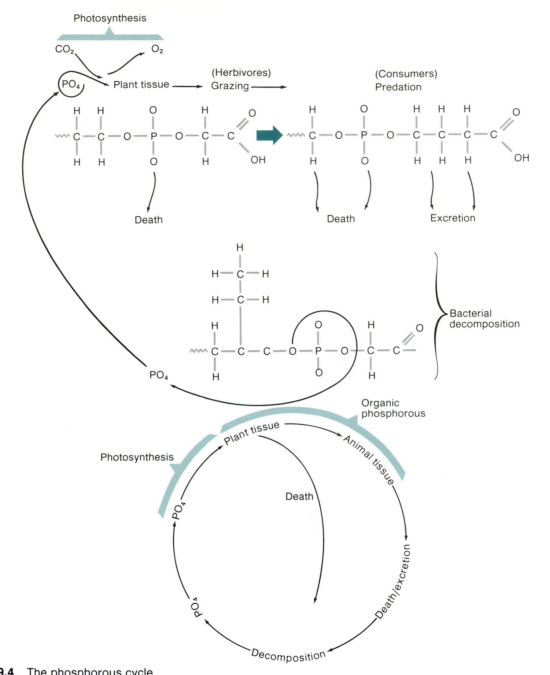

Figure 9.4   The phosphorous cycle

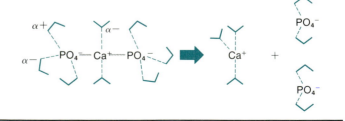

Figure 9.5   The dissociation of calcium phosphate. Since water is polar the partially negative oxygen orients to the calcium, while the partially positive hydrogen surrounds the negative phosphate. Each phosphate breaks away with a calcium electron and becomes a negative ion; the calcium, having lost electrons, assumes a positive charge.

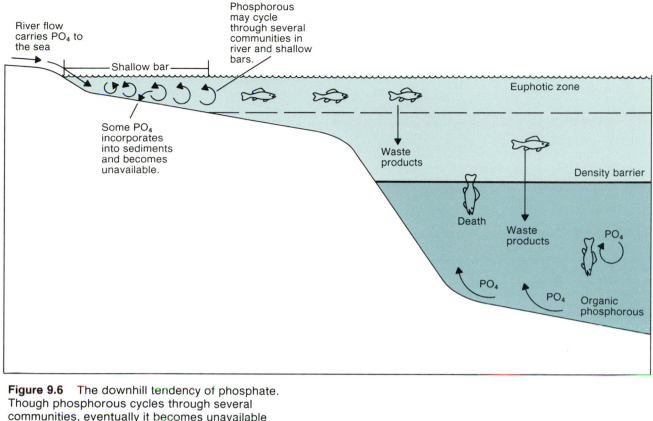

Figure 9.6 The downhill tendency of phosphate. Though phosphorous cycles through several communities, eventually it becomes unavailable when it is incorporated into sediments or is returned as phosphate beneath a density barrier below the euphotic zone.

where it remains chemically unavailable to the plants. Phosphate will also be incorporated into the tissues of marine organisms that inhabit the shallow coastal waters for a time and then move offshore into the deeper ocean. When these organisms die, they sink beneath the euphotic zone. Bacterial decomposition will then return the phosphate to the water beneath the density barriers of the deep ocean, where it will become spatially unavailable for long periods.

Thus, phosphate has a terrestrial origin. Through the chemical weathering of rocks, it dissolves in rainwater and is carried into streams and rivers. In these freshwater systems, it will also cycle from plant to animal to bacteria and back to its inorganic form to be used by other plants farther downstream. In rivers of even moderate length, a phosphate ion may cycle through several, if not hundreds, of communities prior to entering the marine environment.

Upon entering the shallow, coastal sea it may again pass through several communities. Ultimately, however, the phosphate will become incorporated into sediments in an insoluble form or will be carried far offshore. In either case it will become unavailable for long periods of time. This is known as the **downhill tendency of phosphorous** (fig. 9.6).

Nitrogen may be brought into the coastal oceans as dissolved nitrate from terrestrial sources, or it may originate as atmospheric nitrogen gas. When nitrogen gas dissolves into freshwater systems, it will be converted into inorganic nitrate by the blue-green algae that are present in most freshwater environments. The inorganic and organic nitrogen cycle through both freshwater and marine communities in a fashion similar to phosphate, except that nitrate does not tend to chemically combine with other elements to form insoluble compounds; hence, it does not tend to precipitate out in marine sediments as does phosphate.

Nitrate levels are low, however, in the surface waters of the deep temperate ocean, despite the ability of nitrogen gas to diffuse into the water column from the atmosphere. This is due to the absence of the blue-green algae that are found only in the tropical seas.

In coastal areas water from terrestrial sources—rainwater, water used to irrigate farmland, and so forth—often contain relatively high concentrations of dissolved nitrate and phosphate. This water can actually take two different pathways to the sea: it can enter rivers and flow to the sea as surface water, or it can percolate into and through the soil to become groundwater. As groundwater the water will flow "downhill" to the lowest point, as does all water, but at a much slower rate than surface water. Eventually the groundwater will intersect a bay or ocean bottom and enter the marine environment (fig. 9.7). Since the groundwater is less dense than the overlying marine water, it will rise to the surface of the water column.

Dissolved phosphate and nitrate behave very differently when they are carried into the soil. The soil particles tend to develop attractive forces with the phosphate ions and, as a result, the phosphate is retained within the soil column. Thus, phosphate seldom enters into groundwater, nor is it carried into marine systems as a solute. Nitrate and nitrite, on the other hand, do not form attractions with the soils. Consequently these ions readily enter into and travel with the groundwater. As a result of these different interactions, groundwater enters marine systems with varying concentrations of dissolved nitrite and nitrate, but with virtually no dissolved phosphate. This has had significant impacts on coastal waters that are adjacent to moderately and heavily developed uplands or to uplands that are used for agricultural purposes.

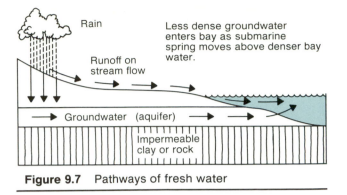

Figure 9.7 Pathways of fresh water

Nitrogen:Phosphorous Ratios

In uncontaminated marine waters, the **nitrogen: phosphorous ratios** tend to be relatively stable. Analysis indicates that for every phosphate there are seven nitrate ions present. Thus, the nitrate phosphate ratio is said to be 7:1. Phytoplankton take up the phosphate and nitrate in this ratio and incorporate it in their tissues. When the grazers consume the phytoplankton, they too take in phosphate and nitrate in these ratios. Then when this organic material decomposes, the nitrate and phosphate are returned to the water in these same ratios. Consequently a normal nitrate to phosphate ratio would be 7:1, regardless of the quantities involved. For example, 1 PO_4 ion to 7 NO_3 ions, 1 milligram of phosphate per liter of water to 7 milligrams of nitrate per liter of water, and 1,000 phosphates to 7,000 nitrates would all maintain the normal 7:1 ratio. Obviously the greater the amount of phosphate and nitrate present in the proper ratios, the larger the population of plants and animals that can be supported, assuming there are proper concentrations of CO_2 and other necessary materials.

Should there be an excess of phosphate—2 milligrams/liter of PO_4 to 7 milligrams/liter of NO_3 for example—the plants would remove all of the nitrate from the water but would only be able to utilize 1 milligram of the phosphate. The remainder could not be utilized, since there is no corresponding amount of nitrate present in the system. In this instance nitrate would be considered to be the ion that would limit plant growth—the limiting factor.

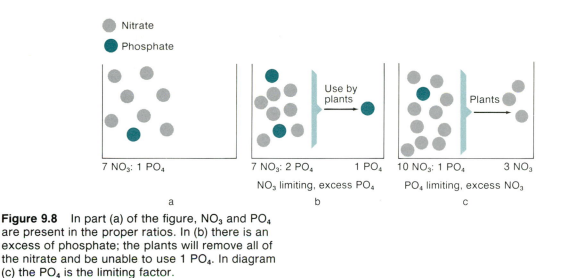

Nitrate

Phosphate

Use by plants

7 NO₃: 1 PO₄

7 NO₃: 2 PO₄   1 PO₄
NO₃ limiting, excess PO₄

10 NO₃: 1 PO₄   3 NO₃
PO₄ limiting, excess NO₃

Plants

a   b   c

Figure 9.8 In part (a) of the figure, NO₃ and PO₄ are present in the proper ratios. In (b) there is an excess of phosphate; the plants will remove all of the nitrate and be unable to use 1 PO₄. In diagram (c) the PO₄ is the limiting factor.

In other cases phosphate could be the limiting factor. For example, if a system contained 1 milligram/liter of phosphate and 10 milligrams/liter of nitrate, the plants would remove all of the phosphate and only be able to utilize 7 milligrams of the nitrate. The remainder would be unconsumed (fig. 9.8).

Nitrogen:Phosphorous Imbalances

Due to the absence of blue-green algae in most of the world's seas, nitrogen tends to be the most common limiting factor in marine systems. In addition most plants inhabiting the seas are adapted to a 1:7 phosphate to nitrate ratio. Plants that require an excess of nitrogen in any form would not be successful in such a system. If present at all, these plants would be in very low numbers and would be unimportant as a food supply. If, however, additional nitrogen entered the system, the limiting factor would be removed and these plants could be expected to proliferate and become more common. Such an imbalance in the normal nitrate:phosphate ratios could have severe consequences for entire communities in marine systems.

A classic example of the consequences of an imbalance in the phosphate:nitrate ratios occurred in the Great South Bay off the south shore of Long Island, New York, in the early 1950s. During this period Long Island experienced a postwar building boom along the western portions of the Great South Bay. To the east, duck farms and other agricultural activity flourished. Human wastes were flushed into subsurface sanitary systems and entered the groundwater. Duck wastes, high in nitrogenous waste products, entered the bay via tributaries, and the inland farms were heavily fertilized and irrigated.

Phosphate, it will be remembered, tends to be attracted by and retained in the soil column, while nitrate and nitrite, as well as other nitrogenous materials, pass freely into the groundwater. Consequently the groundwater showed an increase in the levels of nitrogenous materials. This groundwater, with its high concentration of nitrogenous material, discharged into the Great South Bay. This was augmented by the nitrogen-rich water entering from the duck farms.

The Great South Bay experienced an increase in nitrogen levels in the form of ammonia, urea, and uric acid. This change in both the chemical form of nitrogen, as well as the ratio of nitrogen in relation to phosphorous, led to a change in phytoplankton species.

Previously the bay had been inhabited by populations of diatoms, flagellates, and dinoflagellates. These normal populations were replaced by smaller

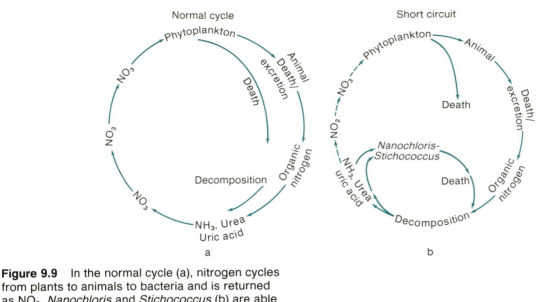

Normal cycle

Short circuit

Figure 9.9 In the normal cycle (a), nitrogen cycles from plants to animals to bacteria and is returned as NO_3. *Nanochloris* and *Stichococcus* (b) are able to utilize nitrogen prior to its conversion to NO_3. Thus the nitrogen, as NO_3, was not available to the normal phytoplankton.

species of flagellates—*Nannochloris* and *Sticho-coccus*. These forms were undigestable by the zoo-plankton. Since both the indigenous phytoplankton and the zooplankton served as the major food supply for the commercially valuable shellfish that inhabited the bay, these species were eliminated by starvation.

Subsequent investigations illustrated that both *Nannochloris* and *Stichococcus* can utilize nitrogen as urea, uric acid, and ammonia. The normal phytoplankton, however, require nitrogen as nitrate. Thus, the new species proliferated and "short-circuited" the nitrogen cycle by removing this nutrient prior to its conversion to nitrate (fig. 9.9). Depriving the normal phytoplankton of their nitrogen source resulted in their elimination from the system. The alteration in the nitrogen:phosphorous ratios led to the proliferation of normally rare forms, which took over a system when it became disturbed. A simple alteration in planktonic species composition led to the virtual elimination of the grazing component of the system, as well as to the loss of a commercially valuable shellfish industry.

The Carbon Cycle

In addition to nitrogen and phosphorous, all organisms require carbon for the production of carbohydrates, fats, and proteins. The atmosphere provides a ready supply of carbon in the form of carbon dioxide (CO_2). Atmospheric carbon dioxide diffuses into the sea at the air–water interface when dissolved CO_2 levels fall below minimum levels.

The **carbon cycle** continues as the dissolved CO_2 is taken in by plants during photosynthesis and is converted into carbohydrates, fats, and proteins. Animals also require carbon but cannot utilize it as carbon dioxide; they are dependent upon the plants for the initial conversion into organic carbon. As a result of grazing and predation, this material enters into the various animal communities and then through excretion and death ultimately reaches the bacteria. The bacteria utilize the material as an energy source and release the CO_2, in its dissolved form, back into the water column. If the water is shallow and well mixed, the CO_2 is rapidly returned to the euphotic zone for reuse by the plants. If it becomes trapped beneath a density barrier and remains spatially unavailable, out of the euphotic zone, additional CO_2 will diffuse into the water column at the air–sea interface. Consequently, carbon depletion is never a problem in the sea, since the atmosphere supplies a continuous supply of carbon in the form of CO_2, which is precisely the form required by plants.

Upon entering the water column, CO_2 can be directly utilized by the plants. It can also be chemically converted into various chemical forms. These conversions are important to maintaining the remarkably constant acid-base levels in the sea. The pH scale is a method devised to determine the acidity of a given system. Waters below a pH of 7 are considered to be acidic, since they have an excess of hydrogen ions (H^+). Above pH 7 waters are basic, since they contain an excess of hydroxyl ions (OH^-).

Depending upon the pH of the system, carbon dioxide can react with water to form carbonic acid (H_2CO_3). The H_2CO_3 may then dissociate to form the bicarbonate ion (HCO_3^{-1}), which, in turn, can further dissociate to form CO_3^{-2}, or the carbonate ion. These dissociations are dependent on the pH of the water column and may be summarized:

$$CO_2 + H_2O \longrightarrow H_2CO_3$$
$$H_2CO_3 \longrightarrow HCO_3^{-1} + H^+$$
$$HCO_3 \longrightarrow CO_3^{-2} + H^+$$

If the marine waters are basic because of an excess of hydroxyl ions, the carbonic acid will dissociate into bicarbonate ions and hydrogen ions. The hydrogen ions will readily react with the hydroxyl ions to form water and thereby reduce the pH.

$$H_2CO_3 + OH^- \longrightarrow HCO_3^{-1} + H-O-H$$

If the waters are acidic due to an excess of hydrogen ions, the carbonate and bicarbonate can accept the excess hydrogen ions, thereby removing them from the water and raising the pH. These reactions may be summarized:

$$HCO_3^{-1} + H^+ \longrightarrow H_2CO_3$$
$$\text{or}$$
$$CO_3^{-2} + H^+ \longrightarrow HCO_3^{-1}$$

The carbonic acid, carbonate, and bicarbonate interactions serve to maintain a constant pH in all marine systems. This phenomenon is known as a **buffer system.** As a result organisms living in the sea are rarely exposed to severe pH fluctuations.

Calcium ions (Ca^{+2}) in the presence of bicarbonate react to form calcium carbonate ($CaCO_3$). $CaCO_3$ is the major component in the shells of mollusks and is also utilized in the formation of coral.

Silicon

Silicon is an essential element for phytoplanktonic diatoms. These forms actively remove silicon from the water and concentrate it in their bodies. Since diatoms are major constituents in phytoplankton communities, silicon is an important factor in the productivity of marine systems.

Silicon originates in terrestrial rocks and sediments and enters the sea in its dissolved ionic form, as well as in undissolved sediment particles. The dissolved silica is removed by diatoms and by grazing and predation passes into the animal component of the system. Through excretion, death, and decomposition, a portion of the silica dissolves, while the remainder becomes incorporated into the bottom sediments. Since silica is continually replaced by terrestrial sediments, no long-term shortages occur. Since, however, it is a trace element that is actively taken up by diatoms during growth and reproduction, short-term shortages do occur. It may be the short-term shortage of silica that limits phytoplankton blooms even when sufficient nitrate, phosphate, and light are present in coastal waters.

Magnesium

Magnesium is required by all plants. This element is an essential constituent of the chlorophyll molecule. Consequently magnesium is actively taken up by all green plants. Magnesium also originates from terrestrial sources and enters the marine system in its dissolved form via river and stream flow. This element passes into the animal portion of the system by grazing and predation and is used in the synthesis of animal enzymes. Through excretion and upon death of the organisms, magnesium is returned by the bacteria during decomposition. Since magnesium, like silica, is continually replaced by terrestrial sources, no long-term shortages occur in coastal areas. However, because of its use in the chlorophyll molecule,

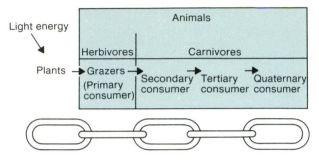

Figure 9.10 A typical food chain

The basking shark feeds exclusively on zooplankton. The zooplankton are filtered by the gill rakers and passed into the stomach while the water is passed out through the gills. Over 4 million tons of water per day is filtered by a basking shark through this process.

it is actively removed from the water and is concentrated in plant tissue. Thus, magnesium levels decline dramatically during plankton blooms. The short-term shortages that occur during these periods may also serve to limit phytoplankton populations.

Food Chains and Trophic Levels

The transfer of energy from the autotrophs to the various heterotrophs can be depicted by **food chains.** The plants form the first link in the food chain, followed by the grazers, the secondary consumers, and so on (fig. 9.10).

Perhaps the simplest food chain is the phytoplankton–zooplankton–basking shark food chain that occurs in the North Atlantic. The basking shark is large, with an average length of 8 meters (25 feet), and feeds exclusively on zooplankton. It filters these organisms from the water by comblike bristles called gill rakers. These feeding structures are arranged to form a mesh size of approximately 0.7 millimeters (0.03 inches), a size slightly smaller than the zooplankton that they feed upon.

When feeding, the basking shark moves at approximately 3 kilometers (2 miles) an hour, with its mouth agape. Water enters its mouth, passes through the gill rakers, and moves out through the gill silts. Passage of water through the gill rakers filters out

Sand eels, like the basking shark, feed on zooplankton.

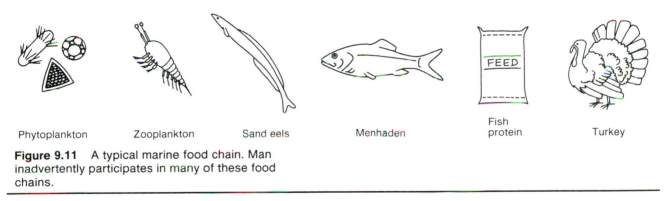

Phytoplankton   Zooplankton   Sand eels   Menhaden   Fish protein   Turkey

Figure 9.11   A typical marine food chain. Man inadvertently participates in many of these food chains.

the zooplankton, which are then passed into the stomach. It has been estimated that a typical basking shark normally filters 200,000 tons of water per hour by this process. Due to the enormous quantity of water that is filtered, the gill rakers wear out and are shed yearly. The shedding occurs in the late summer and early fall after the zooplankton bloom. During this time the basking sharks lie on the bottom without feeding until new gill rakers are grown, having co-ordinated the growth of new gill rakers to coincide with the seasonal decline in the zooplankton populations.

A somewhat longer food chain occurs in the coastal waters of New England. In this case the phytoplankton is fed upon by zooplankton, which, in turn, serves as a food supply for sand eels. The sand eels are fed upon by fish called menhaden. Though the menhaden is considered a trash fish, unfit for human consumption, it is taken commercially and converted into fish protein. The fish protein, or fish flour, is then fed to poultry. Man, by feeding on the poultry, indirectly participates in this food chain as a top consumer (fig. 9.11).

The stage at which an organism obtains its energy is considered to be the **trophic level** occupied by that organism. Since plants obtain their energy directly from the sun, they occupy the first trophic level. The primary consumers occupy the second trophic level, the basking shark and sand eels occupy the third trophic level, the menhaden function at the fourth trophic level, and so on. The final link in all food chains are the bacteria.

Since energy is lost at each link in the food chain, there is an ever-decreasing amount of energy available to the higher trophic levels. Thus, and as dictated by the second law of thermodynamics (chapter 6), there is a definite upper limit to the number of trophic levels and biomass that any given system can support. Basking sharks can survive and prosper in relatively high numbers, since they feed at a low trophic level close to the initial source of energy. Simple food chains, however, tend to be rather unstable, since a loss of any link could have disastrous results for the higher trophic levels. The loss of the zooplankton or sand eels in the menhaden food chain could result in a large loss of menhaden, since few alternate food supplies are available.

In the deep ocean the producers consist solely of the phytoplankton, since sunlight cannot penetrate to the bottom. In the shallow coastal ocean, larger subtidal and intertidal plants are present as well. In these areas the producers consist of the phytoplankton, the sea grass and/or kelp beds and other algae of the subtidal zone, and the salt marshes or mangroves of the intertidal zone.

The kelp, like the phytoplankton, obtains its nutrients directly from the water column. The sea grasses, the plants of the salt marsh, and the mangroves obtain their nutrients from the bottom sediments in which they are rooted. Thus, in these shallow seas the phytoplankton–grazer–secondary and tertiary consumer food chain is present, as well as the more elaborate **detritus** food chains. Detritus is the term given to the small fragments of plant and animal material.

There is very little direct grazing on the plants of the subtidal and intertidal zones. It has been estimated that less than 10 percent of these plants are consumed directly by the grazers. In the marshes of the temperate zone, the majority of the grazing is

The bent-nosed clam, *Macoma sp.,* is a common west coast filter feeder. The clam's ability to survive in stagnant water led to the practice of capturing and keeping these clams in water-filled boxes for several days to flush out any sediment that they may have accumulated prior to marketing them.

carried out by insects that then provide the food supply for many insectivores—birds such as the swallows, martins, and so forth. In these cases the energy removed from the marsh by grazing is diverted to the terrestrial environment by the secondary consumers. Direct grazing is, however, rather insignificant in all cases.

The vast majority of these plants remain unconsumed during the growing season. When the plants die or become dormant, the stems, blades, etc., break off. Currents and tidal action move them about and mechanically reduce them to smaller and smaller fragments. In addition bacterial decomposition converts the cellulose into more readily available and usable carbohydrate. Thus, the plant materials are mechanically reduced to detritus.

Once the detritus is of a small enough size, it will provide a food source for a variety of animals. The detritus that floats in the water is utilized by various filter feeders, while the portion that falls to the bottom serves as a food supply for scavenging crabs and deposit feeders. In addition certain clams—such

Fish occupy different positions in the food chain at different periods in their life cycle.

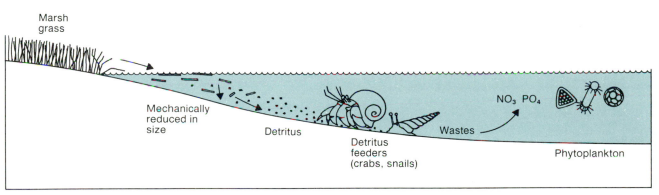

Figure 9.12 The detritus food chain. Ultimately the organisms that feed on detritus excrete wastes that are decomposed, converted into nitrate and phosphate, and utilized by phytoplankton.

as the bent-nosed clam, which is commonly found on mud flats from Kodiak, Alaska, to Baja, California—have developed a specialized apparatus that enables them to siphon detritus from the sediments. These specialized filter feeders are often called suspension feeders. A wide variety of animals, from clams to crabs, feed directly on the detritus. The excretory material released by these organisms is broken down by bacterial decomposition, and the resultant dissolved nitrate and phosphate are then utilized by the phytoplankton. As a result a portion of the nutrients from the macroscopic plants is diverted to the phytoplankton food chain (fig. 9.12).

The major predators on the detritus feeders are fish and birds. The fish tend to occupy different trophic levels at different stages in their life cycles.

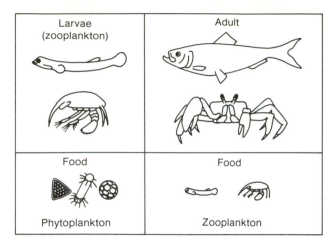

Figure 9.13 Many marine organisms utilize different food sources at different periods in their life cycles.

For example, young larval fish feed on zooplankton and, as they mature, feed directly on detritus. At a later stage they feed on various detritus feeders and ultimately on smaller fish (fig. 9.13). As adults these fish will move offshore, and the materials gleaned from the coastal waters will become spatially unavailable. Since a variety of different species occupy each trophic level in these coastal areas, elaborate food webs (fig. 9.14), rather than simple food chains, develop. Generally food webs are more stable, since alternate food sources exist at each trophic level.

Summary

Biogeochemical cycles involve the transfer of the essential, minor, and trace elements from the abiotic to the biotic components of a system. If essential elements such as the minor element magnesium or light is a limiting factor, all of the phytoplankton in a given area will be affected. If one or more of the other minor elements or the trace elements are limiting factors, only one or two species of phytoplankton will be affected.

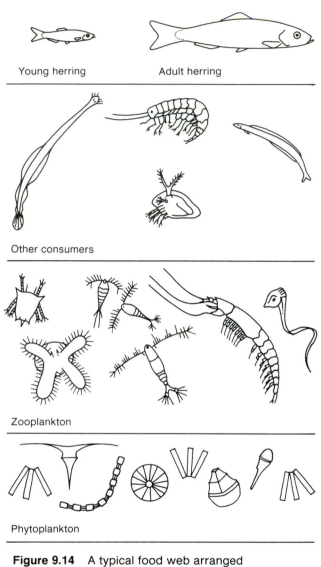

Figure 9.14 A typical food web arranged according to the various trophic levels.

Nitrate and phosphate are present in all natural systems in a 7:1 ratio. If this ratio becomes disrupted, population shifts may occur. Generally these shifts adversely affect the animals that are dependent on the phytoplankton as a food supply.

The essential elements, nitrogen and phosphorous, exhibit a downhill tendency and, as a result, become spatially unavailable for long periods of time. Upwelling guano deposits and fish migrations serve to return a very small fraction of this material to the coastal seas and inland systems.

The transfer of food energy may be depicted by food chains and food webs. Food chains can be and generally are unstable, since the loss of a single link can have drastic effects on the remainder of the chain. Food webs are more stable, since a variety of alternate food sources exist at each trophic level.

Review

1. Distinguish between the essential, minor, and trace elements.
2. Diagram and explain the nitrogen cycle.
3. Diagram and explain the phosphorous cycle.
4. What is meant by the downhill tendency of phosphorous?
5. Why is it important to maintain proper nitrogen:phosphorous ratios in the sea?
6. Explain why light may be limiting during one period and nutrients during another. Give examples.
7. Compare and contrast chemical and spatial unavailability.
8. What is a trophic level? Give examples.
9. Assign each of the following to its appropriate trophic level: basking shark, foraminifera, phytoplankton, copepod, bacteria, herring.
10. Compare and contrast food chains and food webs. Which are more stable? Why?

References

Reisch, D. J. (1969). *Biology of the Oceans*. California: Dickenson Publishing Co.
Russell-Hunter, W. D. (1970). *Aquatic Productivity*. New York: MacMillan.
Steele, J. H. (1970). *Marine Food Chains*. California: U California Press.

For Further Reading

Black, J. A. (1977). *Water Pollution Technology*. Virginia: Reston Publishing Co.
Isaacs, J. D. (1969, September). The nature of oceanic life. *Scientific American*.
Pequegnat, W. E. (1958, January). Whales, plankton and man. *Scientific American*.

10
Winds and Wind-Drift Currents

Key Terms

biogeographic change point
Coriolis effect
Ekman spiral
geostrophic current
high-pressure air mass
low-pressure air mass
three-celled system
upwelling
wind-drift current

Wind-drift currents are formed by wind moving across the surface of the sea and setting water in motion. Winds are formed by the warming and cooling of air masses. This warming and cooling affects the density of the air, causing it to move into adjacent air masses.

The density of air is affected to a greater extent than that of water because the vast majority of the gases that compose the atmosphere are nonpolar, and significant intermolecular forces do not develop between these molecules. Unlike water, air does not require that heat be added to disrupt intermolecular forces in order to increase molecular motion and volume. Since air has a much lower specific heat than water, even small changes in heat cause considerable density changes.

As air is warmed its molecular motion increases; the air occupies a greater volume, becomes less dense, and rises into the atmosphere. When it is cooled the reverse occurs. Molecular motion decreases; the air occupies less volume, becomes denser, and sinks.

The vastly different heat regimes at the equator and the polar regions of the earth's surface are responsible for the great differences in the density of the air masses in these regions. These density differences are intimately involved in the formation of winds and the earth's distinctive patterns of rainfall. The prevailing winds, in turn, drive the larger surface wind-drift currents in all of the world's oceans.

In the deep ocean, wind-drift currents move warm surface water from the equator northward in the Northern Hemisphere and to the south in the Southern Hemisphere. This water is transported for considerable distances and serves to warm the temperate seas of both hemispheres.

Once air or water masses are set in motion, the rotation of the earth influences their pathways. This is most obvious in the surface currents of the deep ocean, where the waters of the Northern Hemisphere circulate from south to north in a large right-handed arc, while the waters of the Southern Hemisphere circulate from north to south in a left-handed arc. As a result the surface currents of both the Atlantic and Pacific Oceans appear to form large figure eights. This deflection of the moving waters of the earth has profound effects on the temperatures of the adjacent landmasses.

Currents formed by the winds may also move water offshore and set up countercurrents that move shoreward. These countercurrents often return nutrients to the euphotic zone by the process of upwelling. In addition winds that blow onshore generate currents that transport sediment along beaches. These currents are responsible for the long-term processes of coastal erosion and deposition.

Air–Sea Relationships: A Simplified Version

If the surface of a large sealed tank of water and overlying air (fig. 10.1) is warmed at the center and cooled at the ends, the warm, central air will increase in volume, become less dense, and rise to a height of equal density. Some of the water in this central portion of the tank will attain sufficient energy to disrupt the intermolecular hydrogen bonds (chapter 7) and break through the surface of the liquid as a warm water vapor. This water vapor will rise with the warm ascending air and, as the air and water vapor remain in contact, the water vapor will impart heat to the air. This will further increase the heat of the air, decrease its density, and result in a well-defined current of rising air containing the water vapor (fig. 10.2).

Since the air in the central portion of the tank is less dense, an area of low pressure will develop. The cooled air at the ends of the tank occupy less volume, are denser, and are under higher pressure; that is, high-pressure systems develop in the colder portions of the tank, while a low-pressure system develops in the central region. The denser high-pressure system will spontaneously move into the less dense low-pressure system over the surface of the water (fig. 10.3). This air will then warm, decrease in density, and rise, etc., just as before.

The warmer central air will move to either end of the tank as an upper air mass (fig. 10.4). If the heat differentials are maintained, the air will then cool, descend, and move to the warmer central portions of the system—the areas of low pressure.

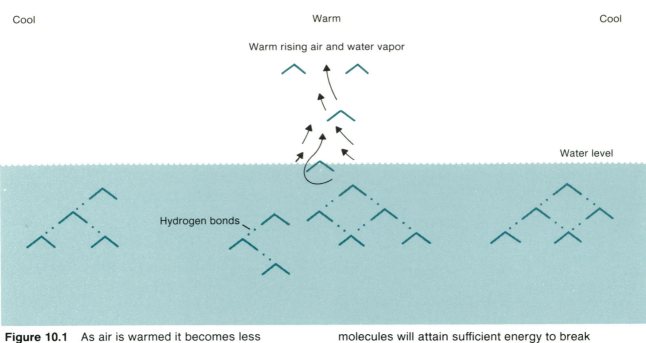

Cool · Warm · Cool

Warm rising air and water vapor

Water level

Hydrogen bonds

Figure 10.1 As air is warmed it becomes less dense and rises. The water will also warm, some of the hydrogen bonds will break, and some of the molecules will attain sufficient energy to break through the surface of the liquid and enter the vapor state.

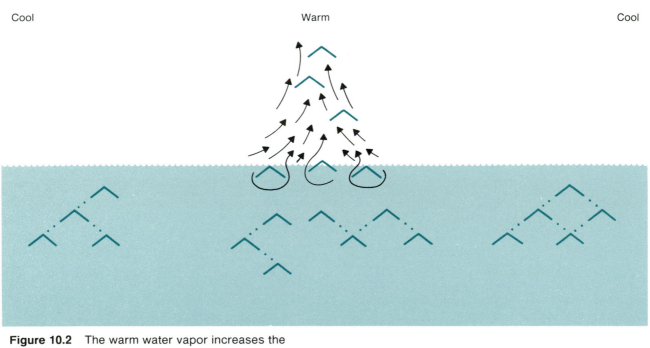

Cool · Warm · Cool

Figure 10.2 The warm water vapor increases the heat of the air in which it is contained. A well-defined current of air and water vapor will rise, since it is less dense than the surrounding atmosphere.

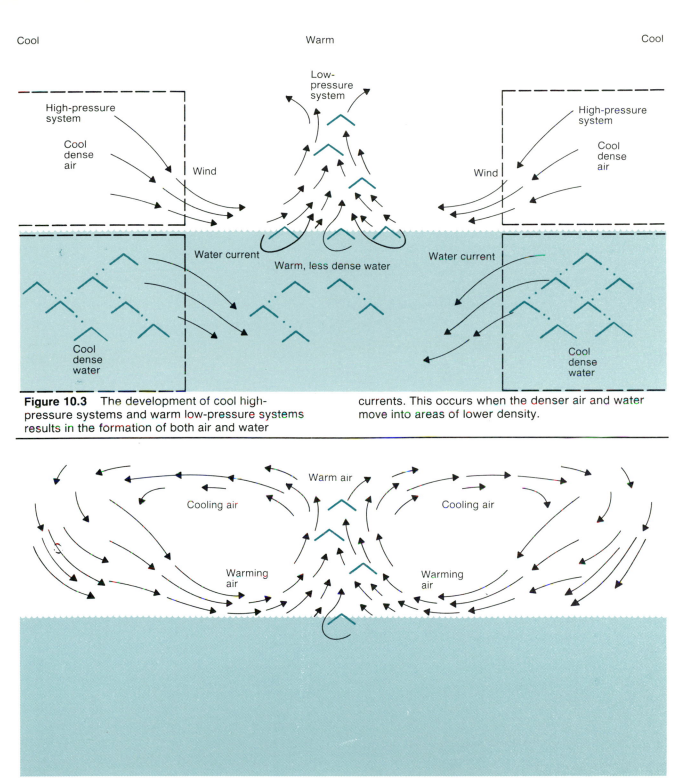

Cool

Warm

Cool

High-pressure
system

Cool
dense
air

Wind

Low-
pressure
system

Water current

Warm, less dense water

Cool
dense
water

Water current

High-pressure
system

Cool
dense
air

Wind

Cool
dense
water

Figure 10.3 The development of cool high-pressure systems and warm low-pressure systems results in the formation of both air and water currents. This occurs when the denser air and water move into areas of lower density.

Cooling air

Warm air

Cooling air

Warming
air

Warming
air

Figure 10.4 As the warm air rises, it will move into colder regions. This air will then cool, increase in density, and become part of and move with the cold air mass. Eventually it will descend and will move over the surface of the water to the central portions of the system, and the process will begin anew.

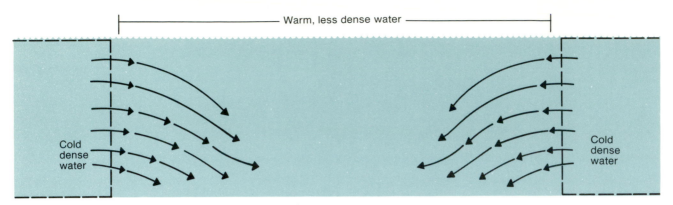

Figure 10.5 A pressure-gradient current is formed when the cold, dense water moves into and sinks beneath the warmer, less dense water.

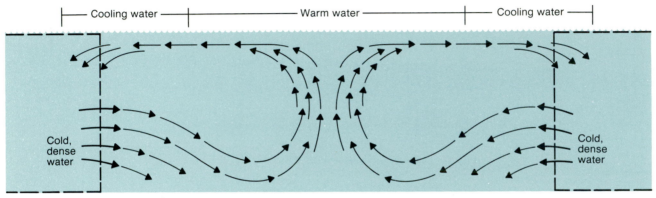

Figure 10.6 The colder, denser water moves into and sinks beneath the warmer, less dense water and then moves toward the central portion of the tank. As it moves into this warmer region, the water will warm, decrease in density, and rise. The warm surface water will move toward the ends of the tank as a surface water current; it will cool and become part of the cold, dense water system.

As a system of warm, rising air is developed in the warmer central portions of the tank, this air is replaced by cooler air that moves over the water's surface from each end of the tank. In other words air under higher pressure will invariably move into and displace air of lower pressure.

The colder water at the ends of the tank will also increase in density and move into and sink beneath the warmer, less dense central water (fig. 10.5). As this subsurface water moves to the central portions of the tank, it will warm, decrease in density, and rise, beginning the cycle anew (fig. 10.6).

Air–Sea Relationships: A Global View

The heat differentials described for the tank are essentially those found on earth. Because of the differences in solar heating (chapter 7) at different areas of the earth's surface, cold water and air are generated in the polar regions, and warm air and water are formed at the equator (fig. 10.7).

As equatorial water is warmed, its evaporation rate increases, and the warm water vapor enters the atmosphere. The overlying air is also warmed, its heat content augmented by that of the water vapor. As a result the air, with its associated water vapor, has a low density and rises to an altitude of equally low density. Once aloft it then moves to the north and

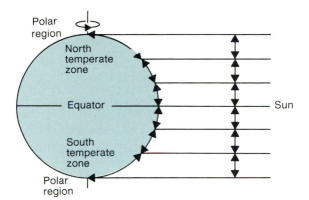

Figure 10.7 The sun's rays strike the equatorial regions most directly, resulting in a greater solar heating of this region of the earth.

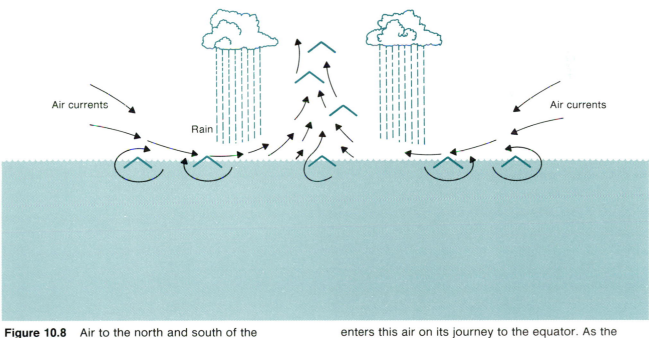

Figure 10.8 Air to the north and south of the equator is cool and dense and moves toward the equator directly over the sea's surface. Water vapor enters this air on its journey to the equator. As the warm air and water vapor rises it cools and the water condenses and falls to earth.

south of the equator. As this air moves away from the equator, it begins to cool and forms hydrogen bonds between the water-vapor molecules, which eventually condense and return to earth as precipitation.

The rising equatorial air mass creates an area of low pressure. Cooler, denser air from the north and south, which is under higher pressure, moves into the equatorial regions as a surface air current moving directly over the water's surface (fig. 10.8). As the

air current moves, it picks up water vapor, which is augmented by additional water vapor at the equator. This air then continues to warm, decreases in density, rises into the atmosphere, and so on.

Due to these interactions the air at the equatorial regions of the earth's surface is warm and has a low density and low atmospheric pressure. The equatorial regions are characterized by clouds and a high precipitation rate. Conversely the temperatures are low and the air denser in the polar regions. The poles have a high atmospheric pressure and the atmosphere contains little water vapor, resulting in less precipitation.

These circulation patterns are similar, but not identical, to those actually observed on earth. Deviations from these "ideal" patterns are due to the rotation of the earth on its axis.

Air–Sea Relationships on a Rotating Earth

The systems described to this point are considered to be single-celled systems. A single column of warm, low-density air rises at the equator or at the central portions of the tank. This air then moves north or south to the poles—or to the ends of the tank—cools, sinks, and returns to the point of initial warming.

In reality the earth's atmosphere behaves as a **three-celled system** due to the rotation of the earth. As the earth rotates from west to east on its axis, different regions are moving at different speeds, since their circumferences differ. The circumference of the earth is the greatest at the equator—the low latitudes—and decreases progressively toward the poles. In fact the earth's speed is calculated to be approximately 1,700 kilometers (1057 miles) per hour at the equator, and only 850 kilometers (528 miles) per hour at latitudes of 60° north and 60° south.

All stationary objects on the earth at these latitudes are actually moving eastward at these speeds. The surrounding atmosphere, attracted to the surface of the earth by gravity, also moves eastward but behaves somewhat more independently due to the low degree of friction between it and the earth's surface.

Should a stationary air mass located at the equator begin to move northward at 15 kilometers (9 miles) per hour, it would have two component motions—eastward at the speed of the earth at the equator and northward at 15 kilometers per hour. As the air continues northward, the circumference of the earth decreases and so too does the earth's speed and the speed of its atmospheric envelope. The air mass is still moving eastward at its original speed of 1,700 kilometers per hour, however. The speed of

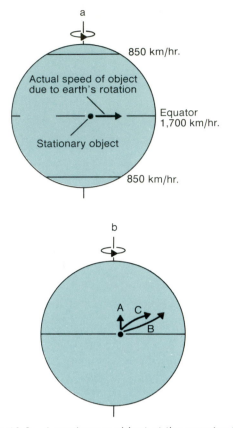

Figure 10.9 A stationary object at the equator (a) would actually be moving eastward at 1,700 km/hr (1057 mph). Diagram (b) shows the object being set in motion northward at 15 km/hr (9 mph) toward point A. Since it is already moving eastward at 1,700 km/hr, it would appear to curve toward its right (C) to an observer on earth. From space, however, it would be apparent that the object is actually moving northeastward in a straight line toward point B.

the air continues to be 1,700 kilometers per hour eastward and 15 kilometers per hour to the north, causing the air mass to move eastward at a more rapid rate than the earth does at these higher latitudes. As a result the air mass will appear to curve to its right (fig. 10.9). Similarly if it is set into motion from the equator southward, it will move into slower-moving latitudes and will appear to curve to its left (fig. 10.10).

Should an air mass be set into motion southward from the North Pole, it will travel into latitudes that are moving at a greater rate of speed, since the circumference of the earth increases toward the

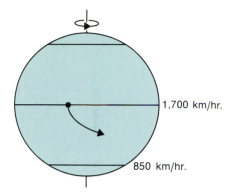

Figure 10.10 Objects moving southward from the equator will also appear to curve—but to their left.

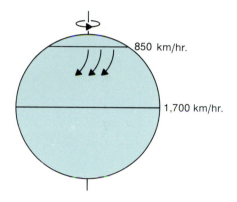

Figure 10.11 Since the earth is moving most rapidly at the equator, an air mass, or any object, that is set in motion southward from the Arctic will appear to curve to its right, since the earth moves out from beneath it.

equator. At these latitudes the earth will move out from beneath the air, which will appear to curve to its right (fig. 10.11). An air mass traveling north from the South Pole will be subjected to a similar set of circumstances and will arc to its left (fig. 10.12). The tendency of moving air masses to curve to their left in the Southern Hemisphere and to their right in the Northern Hemisphere is termed the **Coriolis effect.** The Coriolis effect influences the direction of all moving objects and profoundly affects water currents.

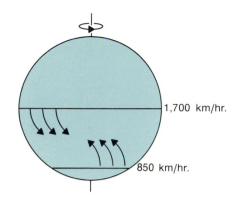

Figure 10.12 Objects moving northward or southward in the Southern Hemisphere will appear to curve to their left.

The Three-Celled Atmospheric Circulation

As a result of the deflections caused by a rotating earth, the air that rises at the equator moves northward in the Northern Hemisphere and begins to curve to its right. As this air moves northward, a portion cools, sinks, and returns to earth in the temperate zone. This descending air mass then travels to the north and south. The remainder of the upper air mass that originated at the equator reaches the polar latitudes, cools, sinks, and moves southward over the sea's surface. Since the air mass is colder than the sea, it is warmed by the sea as it moves south. By the time the air mass has reached the subpolar latitudes of 60° north, it has warmed and decreased in density to the extent that it again rises. As these air masses move, they are influenced by the Coriolis effect and curve to their right. There is a similar movement of air in the Southern Hemisphere; in this case however, the air masses curve to their left. In any event the result of the movement of these air masses over the rotating earth is a three-celled system in each hemisphere.

When the air mass that descends in the temperate zone travels toward the equator, it is also deflected by the Coriolis effect: in the Northern

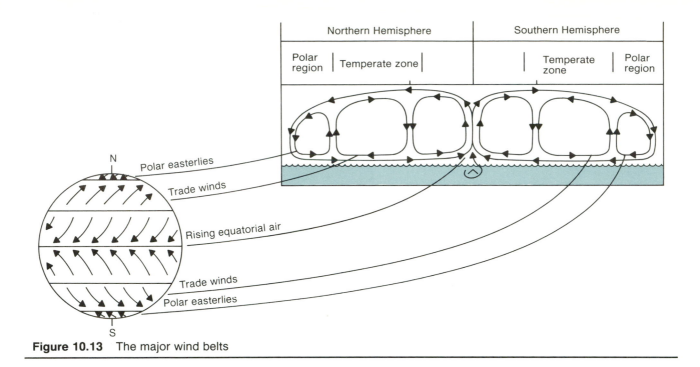

Figure 10.13 The major wind belts

Hemisphere, to its right and in the Southern Hemisphere, to its left. In either case these air masses are deflected to the west and travel in a southwesterly and northwesterly direction, respectively (fig. 10.13). These air masses are known as the trade winds. Since winds are named for the direction from which they blow, the trade winds in the Northern Hemisphere are termed the northeast trade winds, while those in the Southern Hemisphere are called the southeast trade winds.

When this descending air mass travels into the higher latitudes, it is also deflected to its right in the Northern Hemisphere and to its left in the Southern Hemisphere. In these instances the air masses are deflected to the east and travel in a northeasterly and southeasterly direction, respectively. These winds are known as the prevailing westerlies, since they blow from a westerly direction. At the poles the air masses that descend blow toward the equator from the northeast and the southeast and are called the polar easterlies. In addition to the effects exerted by the rotation of the earth, air masses are also influenced by seasonal temperature changes and the continental landmasses.

Box 10.1 The Doldrums and Horse Latitudes

The trade winds, westerlies, and polar easterlies blow consistently from a given direction. These are known as wind belts, or bands. Between the wind belts the surface winds are light and variable. The doldrums are the result of light, variable winds at the equator that are due to the rising air masses in that area.

Other areas of light, variable winds between the major wind bands are found at 30° north and 30° south latitude. These regions are known as the horse latitudes. The name is said to have originated when ships, carrying horses, were becalmed in these latitudes. Supposedly the horses were thrown overboard when drinking water became scarce. At the time the sea was said to be littered with dead and dying horses.

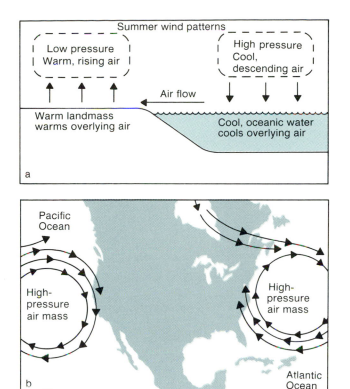

Figure 10.14 As less dense air masses form over the continents in the summer, they move offshore, encounter denser oceanic air, and are forced back over the continent. These air masses are deflected to their right and move in a clockwise direction. As a result the east coast of North America (which is to the west of the oceanic high-pressure system) is warmed by winds from the south, while the west coast (which is to the east of the oceanic high-pressure system) is cooled by northerly winds (b).

The Earth's Atmospheric Circulation

Both the land and the sea are continually cold in the polar regions and warm at the equator and in the tropics. The temperate zones undergo seasonal temperature changes, which profoundly affect the circulation patterns of the atmosphere.

The sea has a much higher specific heat and heat capacity (chapters 6, 7) than either the land or the overlying air; consequently, the oceans warm up and cool down much more slowly. For this reason, in the temperate zone the land is warmer than the sea in the summer and colder in the winter.

In the summer the warm land imparts heat to the overlying air, which decreases in density, rises, and forms a low-pressure area. Simultaneously the air over the sea cools and becomes denser, thereby forming a high-pressure area. The low-pressure areas over the land tend to spread toward each other and move southward toward the equator, while the high-pressure areas are confined to the air over the temperate ocean.

As the low-pressure air cools, it becomes denser and descends over the ocean, where it encounters still denser oceanic air that pushes it back toward the low-pressure air over the continents. As this air moves toward the continents, it is deflected to its right and forms winds that move in a clockwise direction around the **high-pressure air masses** (fig. 10.14). To the north of these high-pressure areas, the winds blow from the west (the westerlies), while to the south they blow from the east. The landmass to the east of each high-pressure system will therefore experience cool winds from the north, while those to the west will be warmed by winds from the south. As a consequence the east coast of the United States receives warm, moist air from the south, while the Pacific coast is cooled by northerly winds in the summer.

In the winter the reverse occurs and high-pressure areas develop over the colder landmasses, while **low-pressure air masses** form over the ocean. This results in a counterclockwise circulation pattern, with a reversal of the prevailing wind patterns (fig. 10.15). During this season the Pacific coast is warmed by southerly winds, while the Atlantic coast is subject to cold air that moves down from the north.

Wind and the Surface Currents of the Sea

When wind blows across the surface of the sea, it sets water into motion as **wind-drift currents**. Depending upon the strength and constancy of the wind, these currents may persist and become a permanent part of the sea's surface circulation pattern, or they may be ill-defined and transitory.

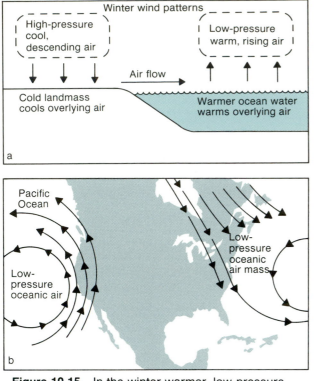

Figure 10.15 In the winter warmer, low-pressure air forms over the ocean, while cold, high-pressure air forms over the land (a). The wind direction shifts in response to these temperature and pressure regimes and the east coast is cooled by cold, northerly air while the west coast is warmed by southerly air (b).

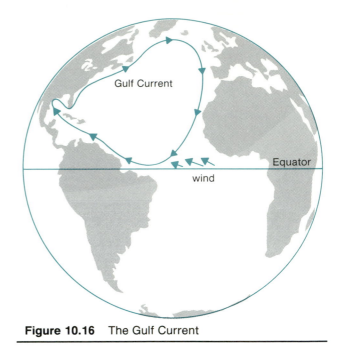

Figure 10.16 The Gulf Current

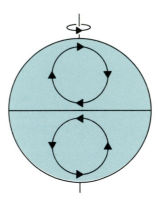

Figure 10.17 Wind-drift currents are formed at the equator and move surface waters northward in a right-handed arc in the Northern Hemisphere and southward in a left-handed arc in the Southern Hemisphere.

Should the direction of the wind remain constant for a long period of time, a considerable amount of water will be set into motion and a well-defined wind-drift current will be formed and persist. The Gulf Stream, or Gulf Current, is such a wind-drift current. This current is set in motion by the prevailing trade winds that move the waters of the Atlantic Ocean in the equatorial region in a north-westerly direction along the coasts of South and North America. This current ultimately circles the entire Atlantic Ocean in the Northern Hemisphere before returning to the equatorial region (fig. 10.16). Similar currents are formed in the Pacific Ocean in both the Northern and Southern Hemispheres, as well as in the Atlantic Ocean in the Southern Hemisphere

(fig. 10.17). These wind-drift currents form the major surface circulation patterns in the world's oceans, and all are generated by the winds that form as a result of the warming of air masses at the sea surface in the equatorial and temperate regions.

Other wind-drift currents are less significant. Variable winds result in ill-defined, and often conflicting, wind-drift currents and current patterns. This

Wind-induced currents that are important in localized coastal areas are known as longshore currents, also called along-shore currents (chapter 5). These currents are capable of transporting tremendous amounts of sediment. For example, a longshore current off the south shore of Long Island, New York, has been estimated to move 45,000 cubic meters (600,000 cubic yards) of sediment westward per year. When longshore currents enter an inlet, harbor, etc., they will disperse and lose velocity. As a result, their carrying capacity is reduced, which results in the deposition of sediment in these areas. Thus, the longshore current is often responsible for causing navigational hazards.

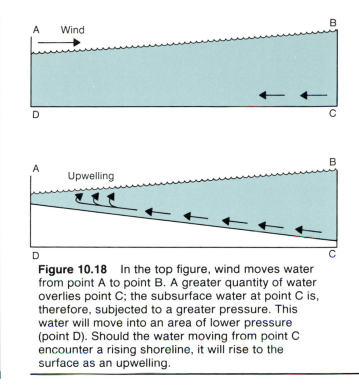

Figure 10.18 In the top figure, wind moves water from point A to point B. A greater quantity of water overlies point C; the subsurface water at point C is, therefore, subjected to a greater pressure. This water will move into an area of lower pressure (point D). Should the water moving from point C encounter a rising shoreline, it will rise to the surface as an upwelling.

is commonly observed in coastal bays, where adjacent land masses deflect winds and cause them to travel over the water surface in a variety of different directions during any given period.

Upwelling is generally the result of both wind-drift and pressure-gradient currents (chapters 7, 11). The wind initiates the process by moving surface waters into a given area. As these waters accumulate, the pressure on the subsurface water is increased. This deep water then begins to move into adjacent waters that are under less pressure (fig. 10.18). Should this deep water encounter a rising shoreline, it will rise to the surface, causing an upwelling.

Another type of upwelling may occur when surface waters move over stationary subsurface water layers. When this happens friction causes the deeper water to begin to move, frequently resulting in the formation of well-defined subsurface water movements, which are then deflected by the Coriolis effect. The movement of this subsurface water sets even deeper water in motion. Since this water is also deflected, this process can result in causing the deepest subsurface water to move in the opposite direction. Should the wind move the surface waters offshore, this reversal may serve to bring subsurface waters onshore as an upwelling. This movement of water is termed the **Ekman spiral** and is discussed in more detail in a later section.

In both types of upwelling, the subsurface water may have been out of the euphotic zone for considerable periods and is consequently high in nutrients. When subsurface water is brought to the surface in this manner, the nutrients become spatially available, which often leads to an increase in plankton populations.

The Coriolis Effect and Water Movements

The Coriolis effect can be observed in the sea, as well as in air masses. In the sea it is best seen in the patterns of the surface currents that travel throughout the deep ocean. It will be recalled that the major surface currents in both the Atlantic and Pacific Oceans appear to travel in large figure eights. The waters of the Northern Hemisphere move in a right-handed arc from the equator to the polar seas and

Box 10.3 El Niño—The Child

Along the coast of Peru, the trade winds move coastal water offshore, creating a countercurrent that brings cold, nutrient-rich subsurface waters onshore where they upwell. This return of nutrients to the euphotic zone brings about a large phytoplankton population, consisting primarily of diatoms. The phytoplankton, in turn, support large populations of copepods, fish larvae, and other primary consumers. The anchovy is the major secondary consumer and it then serves as a food supply for larger fish, squid, and several species of marine birds. As a result of this return of nutrients, these waters are highly productive and account for approximately 20 percent of the world's commercial fishery.

In December the prevailing winds become variable and weaken, ending the upwelling. Since this commonly occurs around Christmas, it is termed El Niño—the Child. With the halt of the upwelling, the temperatures of the surface waters increase and the nutrients decrease. When this occurs the anchovy moves offshore into colder, deeper water, and the tertiary consumers disperse.

This highly productive fishery has been threatened by a series of unfortunate events. In 1972 a prolonged El Niño coincided with an overestimate of the anchovy population. This overestimate led to an increase in the permissible catch limits. As a result the anchovy population was severely overfished at just the wrong time—during a prolonged El Niño—and the catch declined from 20 million metric tons to 2 metric tons in a single year. At present it is doubtful that the anchovy population will recover from these miscalculations.

Another remarkable El Niño occurred during the 1982–1983 season. This El Niño was also prolonged and resulted in abnormally high surface water temperatures. In addition the circulation patterns in the upper atmosphere were affected, which produced the severe storms that struck coastal California.

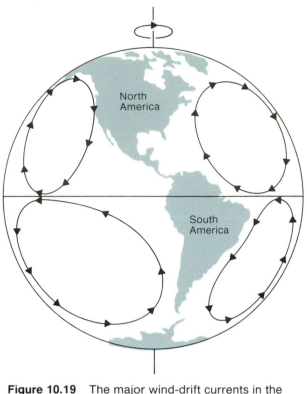

Figure 10.19 The major wind-drift currents in the world's oceans appear to travel in large figure eights due to the Coriolis effect. It should be noted that there is a minimal mixing of water at the equator.

back to the equatorial regions (fig. 10.19). The tendency of water to appear to curve to its right in the Northern Hemisphere and to its left in the Southern Hemisphere is also due to the earth's rotation. The effect is somewhat less pronounced in the sea, however, since there is a greater friction developed between the moving surface waters and the more stationary subsurface waters. As a result there is less of a deflection as these waters travel into different latitudes.

It is the Coriolis effect that determines the pathway of the Gulf Current, which travels in a right-handed arc in the Atlantic Ocean northward from the equator, and the Brazil Current, which travels in a left-handed curve in the Atlantic southward from the equator (fig. 10.20). Similar surface currents also occur in the Pacific Ocean (fig. 10.21). Thus water traveling from the equator northward in the Northern Hemisphere moves in a clockwise direction, while water moving from the equator southward in the Southern Hemisphere moves in a counterclockwise direction.

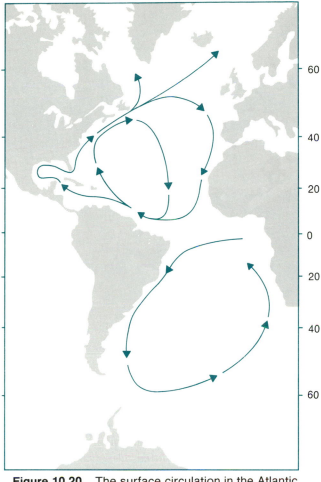

Figure 10.20 The surface circulation in the Atlantic Ocean

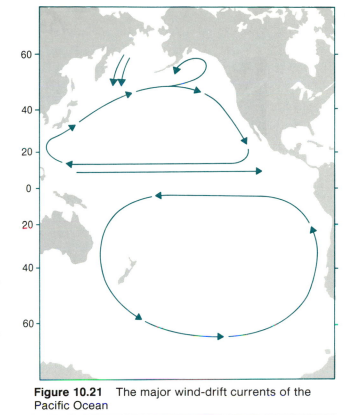

Figure 10.21 The major wind-drift currents of the Pacific Ocean

Since the Coriolis effect moves water in a clockwise direction in the Northern Hemisphere, this water tends to move in a spiral, termed a gyre, and water molecules accumulate within each gyre. This accumulation of water actually causes a small rise in the water levels in these areas. In other words the deflection of water by the Coriolis effect causes the water molecules to form high spots, or hills, on the ocean's surface. Since the earth rotates from west to east, the highest part of each mound is offset to the western boundary of each gyre.

The Coriolis effect continues to add water to these mounds until gravity equals accumulation. When this occurs the water molecules slide down the hill and, as they move downward, come under the influence of and are deflected by the Coriolis effect. The net result is a slow downward motion of the water, with the major motion being parallel to

the sides of the mound. The combined water motion, due to gravity and the Coriolis effect, causes the closed circulation patterns that are common to all the major gyres in the sea. Water movements of this type are called **geostrophic currents.**

The Ekman Spiral

The Coriolis effect causes similar deflection in deeper water as the moving surface water imparts energy to and sets the subsurface water in motion. There is, however, a continual loss of energy because of the relatively small amount of friction that is developed between the water masses. As a result each succeeding subsurface water layer will move more slowly. In the Northern Hemisphere there is a continual deflection of each deeper water mass to the right. The change in current speed and direction with depth would, theoretically, result in a spiral, with the bottommost layer moving in exactly the opposite direction from the wind-drift surface current that began the process (fig. 10.22). A complete reversal rarely, if ever, occurs, since there is a loss of energy as each deeper layer is set into motion. In the

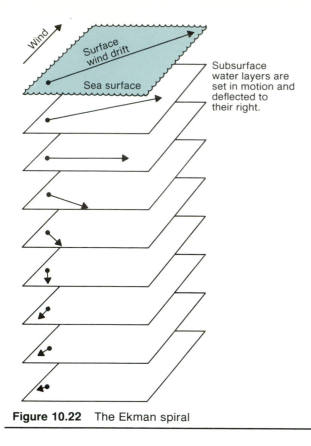

Figure 10.22 The Ekman spiral

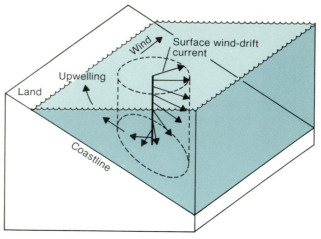

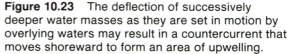

Figure 10.23 The deflection of successively deeper water masses as they are set in motion by overlying waters may result in a countercurrent that moves shoreward to form an area of upwelling.

Southern Hemisphere the water deflects to the left with increasing depth. This phenomenon is termed the Ekman spiral, named for the physicist, Walfrid Ekman, who demonstrated these reversals mathematically.

Although complete reversals rarely occur, the deflection of subsurface water is a major factor in bringing nutrient-rich subsurface water into the euphotic zone in coastal areas. This occurs when the prevailing winds move the surface waters offshore (fig. 10.23). As these wind-driven waters move over and set the subsurface waters into motion, they deflect to their right in the Northern Hemisphere. This water sets deeper water in motion, and it too deflects to its right. The net result forms a subsurface water current traveling shoreward. Should this current encounter and move up a rising coast, nutrients are returned to the euphotic zone, become spatially available, and result in high phytoplankton populations. This is a major factor in the high productivity of many coastal areas.

Wind-Drift Currents

Wind-drift currents form the horizontal surface water movements in the world's oceans. Perhaps the most thoroughly studied of these is the Gulf Current in the Northern Hemisphere (fig. 10.24). The Gulf Current originates at the equator as a result of the trade winds, which set the water in motion toward the coast of South America. This water mass, known as the South Equatorial Current, encounters the bulge of South America and splits into two components. These are the North Equatorial Current, which forms the Gulf Current, and the South Equatorial Current, which forms its counterpart in the Southern Hemisphere.

The North Equatorial Current moves parallel to the equator and then moves northward up the coast of South America. In the vicinity of the Antilles Islands, this water mass splits into the Antilles Current and the Caribbean Current. The Caribbean Current flows between the mainland and the Antilles Islands, while the Antilles Current moves to the east of the Antilles Islands and begins to curve to its right.

The Caribbean Current continues northward as a coastal current and passes over the Yucatan sill and into the Gulf of Mexico. The constricted exit from the Gulf makes the water accumulate in this region as additional water is carried in by the wind-drift Caribbean Current. A relatively high evaporation rate increases the salinity of the water. Ultimately the

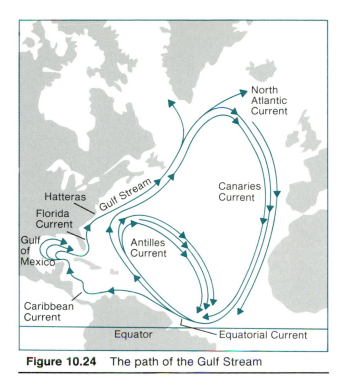

Figure 10.24 The path of the Gulf Stream

water is forced out of the Gulf through the Straits of Florida. At this point it is known as the Florida Current. It has been estimated that 26 million cubic meters of water per second move through the Straits of Florida.

After the Florida Current rounds the tip of Florida, it becomes known as the Gulf Current. As it continues northward, it moves alongside the Antilles Current. This water mass has also been subjected to a high evaporation rate, is very saline, and is denser than the Gulf Current, which received freshwater input from the Mississippi River during its residence in the Gulf of Mexico. As a result of the different densities, a pressure-gradient current develops. Since the water in the Antilles Current is denser, water moves from that current into the Gulf Current, increasing the water volume to 38 million cubic meters per second.

The Gulf Current continues northward as a coastal current. At Cape Hatteras, North Carolina, a combination of prevailing offshore winds, the Hatteras landmass, and the Coriolis effect deflect the water offshore. To the north of Cape Hatteras, the Gulf Current moves past the Sargasso Sea. This water mass is virtually nonmoving and, due to the warm atmospheric temperatures, is very saline. As a result

of the gradual sinking of the surface waters, the highly saline subsurface waters are very dense and contain high concentrations of spatially unavailable nutrients. Since this dense bottom water is subjected to high pressures, it moves into the Gulf Current as a pressure-gradient current. This input of water from the Sargasso Sea increases the transport of water to 55 million cubic meters per second.

The current continues to be called the Gulf Current until it reaches the Grand Banks off Newfoundland. At this point it becomes known as the North Atlantic Current. From here the North Atlantic Current travels eastward across the Atlantic and then diverges to become the Irminger Current, which flows northward along the west coast of Iceland; the Norwegian Current, which moves north along the coast of Norway; and the North Atlantic Current, which completes its right-handed arc and then travels southward along the European coastline. As it passes between the Azores and Spain, it is known as the Canaries Current. This current continues southward to join the Equatorial Current.

The bulge in the South American landmass, which splits the Equatorial Current into two water masses, results in the formation of the Brazil Current in the Southern Hemisphere. This current travels southward along the coast of South America, curves to its left, and travels eastward across the Atlantic. Off the coast of Africa, it merges with the colder Benguela Current and continues northward. In the equatorial region it becomes known as the Southeast Equatorial Current.

It is to be noted that the wind-drift current in the Southern Hemisphere is less complicated than its counterpart in the Northern Hemisphere. This is due to the relatively uniform shoreline in the Southern Hemisphere. In the Northern Hemisphere the current is split by the Antilles Islands, which results in the Caribbean Current's being forced into the Gulf of Mexico and then moving out of the Gulf as a coastal current. This divergence, combined with the residence time in the Gulf of Mexico and the input of warm, saline water from the Antilles Current and the Sargasso Sea, causes the Gulf Current to carry warm water along a large portion of the North American coastline.

Box 10.4 Benjamin Franklin and the Gulph Stream

In the mid-1700s Benjamin Franklin, then colonial postmaster, noticed that mail ships sailing from the colonies to Europe made the trip two weeks faster than those sailing from Europe to North America. This prompted Franklin to ask ship captains to provide him with observations about the surface currents. These observations led him to deduce that the differences in sailing time were due to the current that moved northward along the coast of North America and then eastward across the North Atlantic. In 1770 Franklin had a chart of this current prepared. He also gave instructions as to how to "avoid the Gulph Stream" on voyages from Europe.

The Gulph Stream

Chart courtesy of Philip L. Richardson, Woods Hole Oceanographic Institution, Woods Hole, Massachusetts.

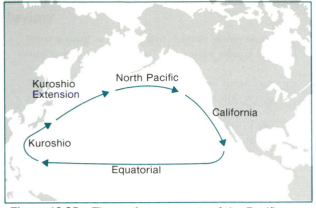

Figure 10.25 The surface currents of the Pacific Ocean in the Northern Hemisphere

The distribution of the southern starfish and northern starfish are influenced by the path of the Gulf Current.

In comparison the warm Brazil Current is diverted offshore rather rapidly by the Coriolis effect. As a consequence it does not moderate the more southerly portions of South America to the extent that the Gulf Current moderates the more northerly portions of North America.

Wind-drift currents also dominate the surface circulation in the Pacific Ocean (fig. 10.25). For example, in the Northern Hemisphere wind sets the water in motion and forms the North Equatorial Current, which initially travels westward across the Pacific Ocean and then curves northward and travels up the coast of Asia. As the water continues northward, it becomes known as the Kuroshio Current. Off the coast of Japan, the Kuroshio Current moves offshore and flows east. At this point it is termed the Kuroshio Extension. This current meets the cold Oyashio Current, which travels south from the Bering Sea. At this point the current, now known as the North Pacific Current, continues to move eastward. A portion of the water in the North Pacific Current continues to move northward to form the Alaskan Gyre, while the remainder diverts southward as the California Current. This current continues south and eventually enters the North Equatorial Current, which is moving to the west at the point of intersection.

The Kuroshio Current, since it travels up the coastline of Japan, may be considered to be the equivalent of that portion of the Gulf Current which travels as a coastal current along the coastline of the southeastern United States. When the Gulf Current diverts offshore at Cape Hatteras it may be considered analogous to the Kuroshio Extension, which also travels northward as a relatively warm offshore water mass.

Biotic Implications

All of these surface currents transport large quantities of warm equatorial and tropical water to the cooler seas. Thus, these currents play an important role in determining the temperature regimes of the world's ocean, which, in turn, influences the distribution of marine organisms. This is particularly true in those areas where the currents move near to, or directly along, the coast.

For example, the waters of the east coast of the United States from Florida to Labrador may be divided into two distinct regions on the basis of water temperature. From the east coast of Florida north to Cape Cod, Massachusetts, the Gulf Current moves either directly along or quite near to the coast. As a consequence these waters are warm enough to allow this area to be classified as the American Atlantic Temperate region. Since, however, the Gulf Current diverts offshore at Cape Hatteras, North Carolina, the water temperatures, and as a result, the biotic assemblages, are sufficiently different to the north and south of Cape Hatteras to warrant subdividing the coastline. Consequently the coastline from Florida to Cape Hatteras is subdivided into the Carolinian province and the coastline from Cape Hatteras to Cape Cod is subdivided into the Virginian province. The region from Cape Cod north to Labrador, where the water temperatures are colder, is considered to be the American Atlantic Boreal region. North of Labrador the water temperatures are polar in nature; hence this area is designated the Atlantic Polar region (fig. 10.26).

Each of these regions and provinces, as noted, has different temperature regimes and, in response, different biotic assemblages. In fact the species composition to the north and south of both Cape Hatteras and Cape Cod is so different that these areas are considered to be **biogeographic change points;** there is an abrupt and obvious change in the biota at these capes. Less than 30 percent of the animals of the Carolinian province are found to the north of Cape Hatteras, while less than 50 percent of the animals of the Boreal region range south of Cape Cod. These temperature regimes, which are brought about by the path of the Gulf Stream, are even reflected in the terrestrial vegetation of the coastal zone. South of Cape Hatteras the woodlands are dominated by the southern long-leaf pine and live oaks, which intergrade with the palmetto in coastal South Carolina. North of Cape Hatteras these species are replaced by the pitch pine and white and black oaks.

The changes in the biotic distribution of marine organisms are due, primarily, to temperature differences over a broad geographical area. In shallow

Figure 10.26 The path of the Gulf Stream directly influences the temperatures of the coastal waters. As a result of the temperature differences, this coastline may be subdivided into biogeographic regions and provinces.

coastal waters such as bays, sounds, and harbors there are also changes in the species composition, but these changes are generally correlated with other abiotic factors, such as salinity and temperature.

Summary

The warming and cooling of air masses sets the atmosphere in motion. These winds, in turn, provide the energy needed to form the wind-drift currents that move throughout the seas of the world.

Once in motion, both air and water masses are influenced by the rotation of the earth. This causes the water masses to appear to deflect to their right in the Northern Hemisphere and to their left in the Southern Hemisphere. As a result the major wind-drift currents appear to travel in large figure eights from the equator northward and southward.

The movement of coastal surface waters offshore may result in the formation of countercurrents by either an increase in the pressure that is exerted on the subsurface waters or by Ekman spiraling. In either case cold, nutrient-rich subsurface waters are brought into the euphotic zone of the coastal areas. This return of nutrients may result in highly productive fisheries in these areas.

Review

1. Explain exactly what occurs when a water and overlying air mass are warmed.
2. Explain how winds are formed.
3. What is a three-celled system?
4. The east coast of the United States is warmer in the summer and colder in the winter than the west coast. Why?
5. Explain how upwellings may occur.
6. What is the Ekman spiral?
7. Sketch the paths of the major wind-drift currents in the Northern and Southern Hemispheres of both the Atlantic and Pacific Oceans.
8. Why do these water masses (question 7) appear to curve?
9. Why are the southerly portions of the east coast of South America colder than the northerly portions of North America at the same latitude?
10. How does the Gulf Current influence the biotic assemblages?

References

Brooks, C. E. P. (1970). *Climate Through the Ages.* New York: Dover.

Miller, A. (1971). *Meteorology.* Ohio: Merrill.

Pickard, G. L. (1963). *Descriptive Physical Oceanography.* New York: Pergamon.

Stommel, H. (1965). *The Gulf Stream: A Physical and Dynamical Description.* California: U of Cal. Press.

Von Arx, W. S. (1962). *An Introduction to Physical Oceanography.* Mass.: Addison-Wesley.

For Further Reading

Anikouchine, W. A., & Sternberg, R. W. (1981). *The World Ocean,* New Jersey: Prentice Hall.

Gross, M. G. 1977. *Oceanography: A View of the Earth,* New Jersey: Prentice Hall.

McDonald, J. E. (1952, May). The coriolis effect. *Scientific American.*

Munk, W. (1955). The circulation of the oceans. *Scientific American.*

11

The Thermohaline Circulation: The Subsurface Currents

Key Terms

advection
Antarctic bottom water
Antarctic source water
Arctic bottom water
Arctic source water
central water mass
circumpolar water mass
contour current
density current
density differential
estuary
euryhaline organism
eurythermal organism
halocline

highly stratified estuary
hypersaline estuary
lesser circulation
moderately stratified estuary
North Atlantic deep water
saltwater wedge
stenohaline organism
stenothermal organism
subpolar water mass
thermohaline circulation
thermohaline currents
turbidity current
vertically homogeneous estuary

Although the surface currents of the sea are the most obvious and move at the greatest speeds, they move through and involve less than 10 percent of the sea's water. The subsurface currents, on the other hand, although they move much more slowly, are difficult to study since, for the most part, they move deep beneath the sea's surface. Yet these currents move through and involve more than 90 percent of the ocean's water volume.

Subsurface water masses can be considered **density currents,** which are induced as a result of temperature and/or salinity differentials that develop between adjacent bodies of water. The denser water will invariably move into and sink beneath the less dense water. Once the dense water reaches a depth of equal density, it will begin to move horizontally. Density currents therefore, have both horizontal and vertical components. Initially they move horizontally as pressure-gradient currents into a water mass of lower density and, hence, pressure; they then move vertically as they sink beneath the less dense water. Upon reaching a depth of equal density, they again move horizontally beneath the less dense water.

Density currents are responsible for the loss of nutrients from the euphotic zone, as well as for the transport of oxygen to the dysphotic zone. Since nutrients are dissolved in very dense water, they tend to remain spatially unavailable for long periods of time and are transported for great distances by these currents.

In shallow inshore areas, tides bring saline oceanic water into bays, where it meets the less dense coastal water. This forms tidally induced density currents, which establish the characteristic salinity patterns of these areas. In addition saltwater wedges may form in rivers and travel far upstream. These salinity differences often influence the distribution of both animals and plants.

Formation of Density Currents

On a small scale density currents can be formed by placing water masses of different salinities and/or temperatures in a partitioned tank. A model akin to the earth's ocean can be constructed by dividing the tank into three compartments, separated by movable partitions. Water with a low salinity and high temperature is placed in the central compartment and water with a high salinity and low temperature is put into each of the end compartments (fig. 11.1).

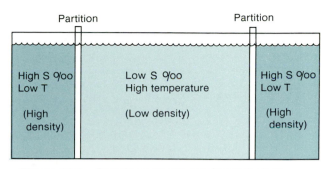

Figure 11.1   Density currents are formed when water masses of different densities are adjacent to each other. In this example the water at the ends of the tank are similar to the dense Arctic and Antarctic water masses. That in the central portion is akin to temperate waters.

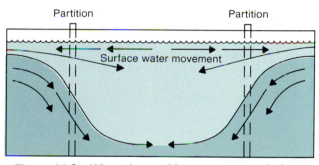

Figure 11.2   When the partitions are removed, the denser water will move into and sink beneath the less dense water.

Since the waters confined at the ends of the tank have a high salinity and low temperature, they are denser and are therefore subjected to a greater pressure than the central water mass. When the partitions are removed, these waters will move spontaneously into the low-pressure area of the tank, moving from top to bottom into the central water. Since they are denser, they will sink beneath the central water and will move as a bottom current toward the opposite ends of the tank (fig. 11.2).

Should one of these bottom-moving water masses be denser than the other, it will, when they meet, sink below the less dense water as it moves to the opposite end of the tank. The less dense water will move above the bottom water mass and travel to the other end of the tank (fig. 11.3). This will result in a three-layered system, with the water masses separated by temperature-density barriers (chapter 7).

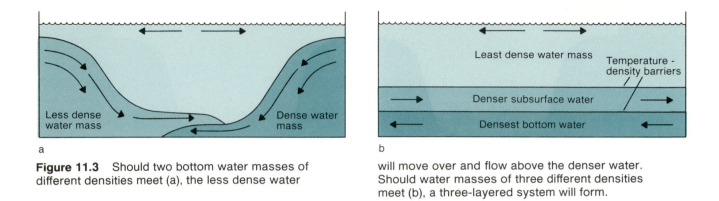

a

b

Figure 11.3 Should two bottom water masses of different densities meet (a), the less dense water will move over and flow above the denser water. Should water masses of three different densities meet (b), a three-layered system will form.

Since these water masses move in response to pressure differentials, they are a type of pressure-gradient current. The pressure differentials, in this case, arise as a result of the differences in the density of the water masses, which are brought about by their different temperature and salinity regimes. Thus, these currents may be termed **thermohaline** (*therm* heat + *hal* salt) **currents** in order to distinguish them from the pressure-gradient currents involved in upwelling (chapter 10). The latter currents are generated by offshore winds that moved water and caused it to accumulate in a given area. The weight of the additional water, rather than the density of the water mass, was the generating factor responsible for those water movements.

Thermohaline currents always move in response to **density differentials,** which arise when a cold and/or highly saline water mass is adjacent to a warmer and/or less saline water mass. The colder or more saline water is denser and therefore under greater pressure than the warmer, less saline water, with the result that the denser water will move into and sink beneath the less dense water.

Surface Water Masses and Thermohaline Circulation

Subsurface currents are generated when pressure differentials develop between adjacent water masses. The pressure differentials are due to the different densities of these water masses, which, in turn, are due to their temperatures and/or salinities. Since the temperature and salinity of water masses are altered by warming and cooling at the surface (chapter 7), it is important to know the characteristics of the surface water masses of the world's ocean.

The surface waters of the sea may be divided into three major water masses: the **central,** the **subpolar,** and the **circumpolar.** They are further subdivided into the northern and southern central, the Arctic and Antarctic subpolar, and the Arctic and Antarctic circumpolar masses, depending upon the hemisphere in which they are found (fig. 11.4).

The central water masses are located in the temperate zones of each hemisphere between latitudes 30° and 45°. These water masses have high salinities and temperatures, ranging between 34–36 ‰ and 6–19°C (43–66°F), respectively.

The subpolar water masses are found in the higher latitudes at 50° north and south. Although they have lower salinity due to the greater rainfall at these latitudes, their temperatures are also markedly colder, ranging from 0°C (32°F) to 2°C (36°F). The colder temperatures cause these waters to be denser than those of the central water masses, even though their salinity is lower.

The circumpolar water masses, as their name implies, are located in the polar seas. Since these waters are very cold, large amounts of sea ice are formed. As the ice forms, dissolved materials are excluded from the ice crystal and remain in solution (chapter 6). This serves to increase the salinity of the unfrozen water and simultaneously depresses its freezing point. As a result the circumpolar waters are the densest in the world.

Mechanisms of the Thermohaline Circulation

The salinity and temperature regimes of the central, subpolar, and circumpolar water masses determine the pattern of the **thermohaline circulation.** The very dense circumpolar water masses move into and

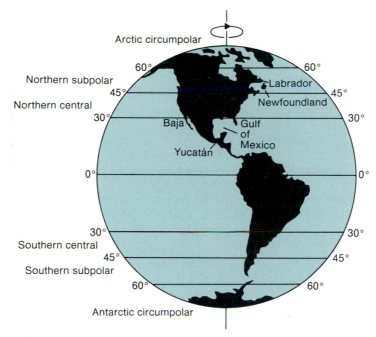

Figure 11.4 The major surface water masses of the sea

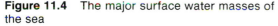

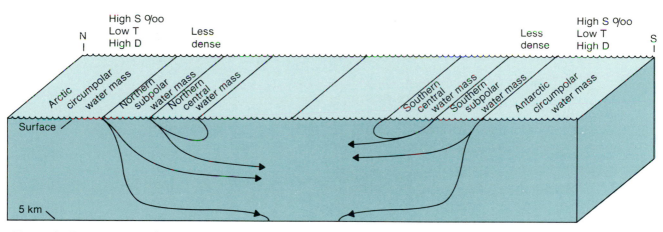

Figure 11.5 The major subsurface water masses. The surface water masses from figure 11.4 are also indicated.

sink beneath the adjacent subpolar water masses in each hemisphere. Since these water masses are the densest in the world, they sink to and travel along the ocean floor.

The waters of the subpolar water masses also move into and sink beneath the less dense central water mass. The subpolar water is less dense than the circumpolar water; consequently, it moves toward the equator as an intermediate water mass in both hemispheres. There are, then, essentially two subsurface water masses in motion in each hemisphere—the intermediate water that originates from the subpolar masses and the bottom waters that originate from the circumpolar water mass (fig. 11.5).

Box 11.1 Sampling The Deep Ocean

Since the different water masses that compose the thermohaline circulation have distinctive salinities and temperatures, it is possible to trace these currents by determining the salinity and temperature of the sea at different depths. This data can then be correlated to the temperature and salinity of the surface waters from which the deeper waters originated.

In sampling subsurface water masses, one encounters the problem of obtaining a water sample from a specific depth and retrieving it without having the sample mix with water from other depths. In addition the temperature cannot be measured after the sample is brought aboard the ship, since the water temperature would change as the sample traveled through various water depths at different temperatures.

The problem is solved by using specially designed sample bottles and thermometers. These water samplers consist of open cylinders that are attached to a weighted wire and lowered to the desired depth. The bottles are equipped with valves and triggers that are activated by a weight, called a messenger. Once the water bottles have been lowered to the desired depth, the messenger is attached to the wire and released. It travels down the wire, strikes the trigger on the water bottle, and closes the valve. When the valve closes, a water sample at that particular depth has been obtained. It can then be returned to the ship for analysis.

The most common sample bottle is the Nansen bottle. The Nansen bottle is attached at its top and bottom to the wire. When the messenger strikes the trigger on the Nansen bottle, the trigger releases the top portion of the bottle from the wire and causes the bottle to release and reverse. As the bottle turns over, the valves close and obtain a water sample. Generally several Nansen bottles are attached to the wire to allow a series of samples to be taken at different depths. As the first messenger strikes the topmost bottle, it reverses and in the process releases another messenger, which then travels down the wire to strike the next Nansen bottle and repeat the process.

Temperatures are taken as the water sample is obtained by using reversing thermometers that are attached to the Nansen bottle. These thermometers consist of a capillary tube drawn out into a loop, termed a pigtail, and a constriction immediately above the mercury reservoir. When the Nansen bottle with the attached thermometer reverses, the mercury column of the thermometer separates from the reservoir. Once it is separated from the reservoir, the mercury in the column will not respond to further temperature changes. Hence, the temperature of the water is recorded at the instant that the water bottle reverses and takes the water sample.

Since cold water generally contains higher concentrations of dissolved oxygen than does warm water (chapter 7), these currents carry oxygen to the deeper water. This process is termed **advection.** Should nutrients be present in these sinking water masses, they will be carried out of the euphotic zone and become spatially unavailable for long periods of time.

The Thermohaline Circulation

The cold atmospheric temperatures in the Arctic and Antarctic regions cause large quantities of sea ice to form. Dissolved materials excluded from these ice crystals remain in solution and increase the salinity of the unfrozen water. As a result the Arctic and Antarctic waters are very cold, highly saline, and very dense. The waters in the Antarctic have salinities as high as 37 %oo and temperatures as low as −2°C (28°F), while the Arctic waters have salinities of approximately 35 %oo and temperatures as low as −0.5°C (31°F). As a result Antarctic water is the densest water in the world's ocean, while the Arctic

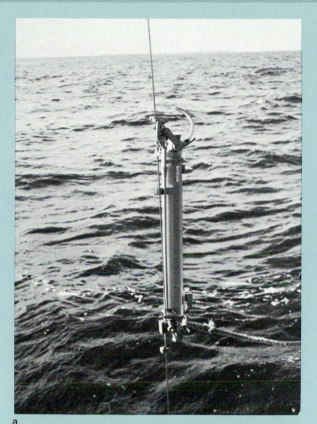

a Nansen Bottles before (a) and after (b)
obtaining a seawater sample. b

waters are only slightly less dense. In order to distinguish between these water masses, they may be termed **Arctic source water** and **Antarctic source water,** since they acquire their characteristic temperature, salinity, and density in these regions.

Both of these water masses move into and sink beneath the less dense, adjacent water masses. Due to their great density, both the Arctic source water and the Antarctic source water sink to the bottom of the sea and move over the ocean floor. The Arctic source water travels south over the sea floor; however, in the vicinity of Greenland it encounters the Arctic sill, which prevents its penetrating any farther southward. The Antarctic source water, on the other hand, travels down the continental slope, moves northward along the ocean bottom and penetrates all the oceans of the world. This water mass is also often referred to as **Antarctic bottom water.** It has been found as far north as 30° north latitude.

The Arctic and Antarctic subpolar water masses are also adjacent to less dense waters. As a result of these density differentials, the subpolar water masses move into and sink beneath the less dense water. They then move toward the equator as intermediate water masses.

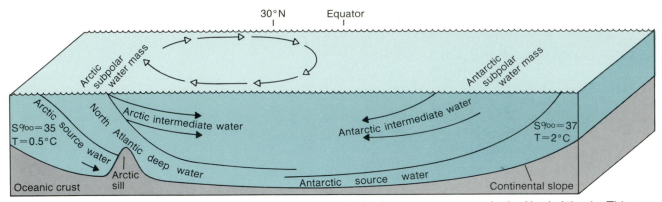

Figure 11.6 The thermohaline circulation. Note that the Arctic source water is trapped behind the Arctic sill and penetrates no farther southward. Thus, south of the sill the North Atlantic deep water is the bottom water mass in the North Atlantic. This water mass is dense, since it receives highly saline water from the Gulf Current. When the North Atlantic deep water encounters the denser Antarctic source water, it moves above it.

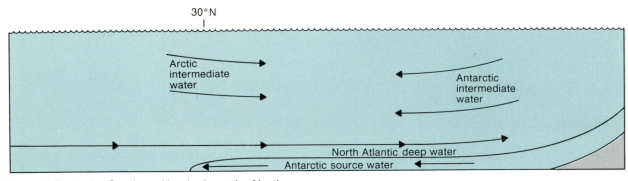

Figure 11.7 In the Southern Hemisphere the North Atlantic deep water is found between the less dense Antarctic intermediate water and the denser Antarctic source water.

Another water mass is also formed in the North Atlantic. Termed **North Atlantic deep water,** it is formed when highly saline water from the Gulf Current mixes with Arctic subpolar water. Because of the input of the saline Gulf water, this water mass is very dense, sinks to the bottom, and travels southward directly over the sea floor. When the denser Antarctic bottom water encounters the North Atlantic deep water at 30° north latitude, the North Atlantic deep water moves above it. The result is that the North Atlantic deep water travels as a bottom current southward to 30° north latitude. Figure 11.6 illustrates the thermohaline circulation. Farther to the south it travels between the **Arctic bottom water** and the upper intermediate water mass (fig. 11.7).

The North Atlantic deep water continues south, eventually encountering the mass of very dense descending Antarctic source water. This cold, dense mass of Antarctic water is very resistant to mixing. As a result the Atlantic deep water is forced upward along its slope to the surface. Since this water has been out of the euphotic zone for hundreds of years, large quantities of nutrients have accumulated. With the water's surfacing in the Antarctic, these nutrients are returned to the euphotic zone, where they bring about a large, fairly continuous phytoplankton bloom. The availability of phytoplankton triggers a zooplankton increase and a concomitant increase in other consumers.

Box 11.2   Other Currents of the Deep Ocean

Turbidity currents and **contour currents** are responsible for shaping the topography of the deep-sea floor (chapter 2). Turbidity currents consist of a mixture of water and sediments, are very dense, and as a consequence travel along the bottom. These currents transport sediment from the continental margins to the continental rises and far out onto the deep-ocean floor in both the Atlantic and the Indian oceans. It is the sediments deposited by turbidity currents that obscure the topography underlying the broad, flat, and featureless abyssal plains that are present in these oceans. In the southern Pacific Ocean, however, the close proximity to the continental margins of the east Pacific rise and the trenches prevents these sediment-laden turbidity currents from traveling onto and depositing sediments on the sea floor farther offshore.

Contour currents flow parallel to the continental slopes at very slow speeds. The contour current that flows along the continental slope in the Atlantic Ocean moves from north to south and is known to contain more water in the portion that flows along the slope and less in the offshore portion. As a result, the landward portion of this current carries, and eventually deposits, a greater amount of sediment than the seaward portion. The sediment is deposited in a wedge shape, with the thinner portion located farther offshore. Thus, contour currents influence the shape of the continental rise.

Contour currents flow parallel to the continental slope. The landward portions of these currents contain more water and therefore move a greater quantity of sediment than do the seaward portions.

Due to the unequal movement of sediments by a contour current, the sediment, when ultimately deposited, forms a wedge shape over the sea floor, with the lesser amount of sediment deposited the farthest offshore the continental slope.

As noted previously the Arctic sill blocks the flow of Arctic bottom water and prevents it from entering the Atlantic Ocean. In the Pacific Ocean, the Bering sill presents an even more efficient barrier. Thus, the majority of Arctic source water is retained in the Arctic seas, and only a very small portion enters into the Atlantic and Pacific oceans.

This massive movement of water away from the Arctic and Antarctic results in a north- and southward flow of warmer water. Warm surface water from the equatorial regions tends to move northward and southward in response to the loss of water from the polar seas. This warm surface water has a relatively high salinity due to the high evaporation rate in the equatorial regions. The high temperatures of this water, however, reduce its density significantly. As this water moves toward the polar regions, it cools, increases in density, and becomes indistinguishable from the surrounding subpolar water masses. This circulation pattern is often known as the **lesser circulation** (fig. 11.8). The lesser circulation is often obscured by wind and wind-drift currents.

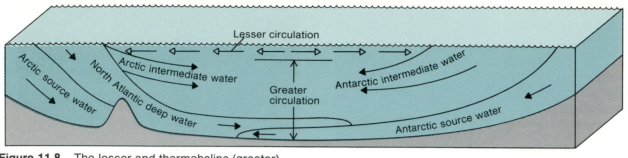

Figure 11.8 The lesser and thermohaline (greater) circulation systems

Estuarine Characteristics and Circulation Patterns

Bays, sounds, etc., are considered to be **estuaries,** which are defined as being semi-enclosed bodies of water with free connections to the ocean. Since estuaries are adjacent to land masses, they receive fresh water and terrestrial sediments from river flow and surface runoff, as well as saline water and marine sediments from the oceanic waters. These areas are the site of the meeting and mixing of oceanic and freshwater systems. Therefore estuaries are considered to be transition zones, or ecotones, between fresh and marine environments.

Currents in estuaries, as in the deep ocean, are generated by both the wind and density differentials. Wind-drift currents are, however, generally transitory and ill-defined, since the adjacent landmasses often deflect the winds and cause them to blow over the surface of the estuary in a variety of different directions during any given period. As a result wind-drift currents traveling in predictable directions seldom have the opportunity to develop in typical estuaries.

Since estuaries receive the input of both fresh and marine water, density-induced currents are generally dominant. Estuaries are often classified on the basis of the distribution of water masses of different densities. According to this method estuaries may be classified as **highly stratified, moderately stratified, vertically homogeneous,** and **hypersaline.** The water in hypersaline estuaries is highly saline and therefore very dense.

Highly stratified estuaries result from the inflow of large quantities of river water. The denser estuarine water encounters the less dense river water,

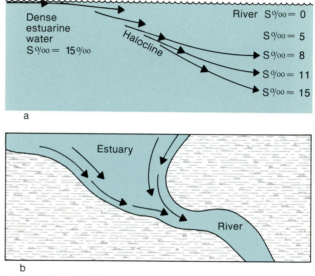

Figure 11.9 A highly stratified estuary. Note that the salinity varies with depth.

sinks beneath it, and travels upstream as a subsurface **saltwater wedge** (fig. 11.9). A salt slope, or **halocline,** is formed at the boundary of the saltwater wedge. A halocline is a water mass that has a large change in salinity from its top to its bottom. In large river systems the wedge is capable of extending for considerable distances upstream, especially during periods of high tide. It is the saltwater wedge in highly stratified estuaries that is responsible for the presence of freshwater organisms in the surface waters and for the typical marine organisms in the waters directly beneath. The mouth of the Hudson River is an example of a highly stratified estuary, and the saltwater wedge is the primary factor involved in the presence of marine organisms far upstream.

The Hudson River is a highly stratified estuary. The estuaries behind the barrier beaches are vertically homogeneous.

Moderately stratified estuaries are the result of essentially equal inputs of fresh and tidal water. In these estuaries mixing is due, primarily, to the turbulence that occurs during tidal changes. This tidal mixing results in a complex pattern of water layers. In these estuaries the halocline is not as steep as in highly stratified estuaries.

During wet periods with increased surface runoff and stream flow, some moderately stratified estuaries will become highly stratified. During these periods the halocline will become steeper and extend farther upstream. In drier seasons surface runoff and stream flow will decline, and the estuary will again become moderately stratified.

Vertically homogeneous estuaries are the result of situations in which the tidal action is dominant. Due to the volume of water brought in at high tide, the water column tends to be well mixed from top to bottom. When the tide is rising, the salinity increases significantly. The salinity and temperature in vertically homogeneous estuaries vary horizontally (fig. 11.10) rather than vertically, as in highly stratified estuaries. This is due to the low volume of stream flow.

Hypersaline estuaries are the result of a combination of low freshwater input and low tidal amplitude, combined with a high evaporation rate; consequently, the salinity in these estuaries is high.

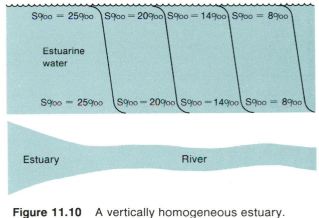

Figure 11.10 A vertically homogeneous estuary. The salinity is the same from top to bottom but varies between the seaward and landward portions.

During very warm periods the salinity in hypersaline estuaries may actually be higher than the salinity of oceanic water.

The Coriolis effect is generally negligible in most estuaries; however, in very wide estuaries the deflection of water becomes obvious. For example, the Chesapeake Bay on the mid-Atlantic coast has a very wide opening to the ocean (fig. 11.11). When marine water enters this estuary on a rising tide, it is deflected to its right along the eastern shoreline.

The Thermohaline Circulation: The Subsurface Currents   199

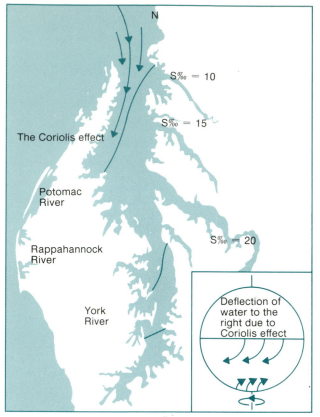

S‰ = 10

S‰ = 15

S‰ = 20

N

The Coriolis effect

Potomac
River

Rappahannock
River

York
River

Deflection of
water to the
right due to
Coriolis effect

Figure 11.11 As sea water enters the Chesapeake Bay, it is deflected to its right by the Coriolis effect (inset). Note the higher salinities of the eastern shoreline. Lines indicate water of equal salinity.

Fresh water traveling downstream is also deflected to its right and accumulates along the western shoreline. As a result the eastern shore of Chesapeake Bay has higher salinities than the western side.

It has already been noted that estuaries receive terrestrial sediments by river flow and surface runoff, as well as marine sediments that enter through the inlets to the sea. The sediments derived from terrestrial sources are generally fine and remain suspended in the water column for considerable periods of time before settling to the bottom of the estuary. While they are suspended these sediments serve to increase the turbidity of the water and thereby decrease the depth of the euphotic zone. This tends to reduce the plant populations and reduces photosynthesis. Generally estuarine water is least turbid in the vicinity of inlets, where oceanic water enters.

Since oceanic water generally moves through an inlet with considerable velocity, it is capable of carrying large quantities of sediment into an estuary. Upon entering the estuary the oceanic water will disperse, its carrying capacity will decrease, and the coarser, heavier sediments will begin to settle out.

As a result of the deposition of both marine and terrestrial sediments, estuaries are rather shallow. Since shallow water is more rapidly heated and cooled than deeper water, estuaries are subjected to rather wide temperature variations. These variations are intensified by the input of water from rivers, streams, surface runoff, and so forth. Since these freshwater systems are comparatively shallow, they are also influenced by atmospheric temperatures. Rivers in the temperate zone therefore tend to be colder in the winter and warmer in the summer than the adjacent marine waters. Consequently when these fresh waters enter an estuary and mix with the sea water, they alter the temperature of the entire water mass, making estuarine waters warmer in the summer and colder in the winter than the surrounding coastal water. Since the fresh water that enters an estuary is less dense than the marine water, it tends to layer over the denser water in many instances. Hence estuarine temperatures often vary vertically.

All these factors make estuaries very variable in terms of salinity and temperature. The temperatures remain relatively constant throughout the entire estuary during any given period of time, but they do vary considerably between summer and winter. Salinity varies with location within the estuary; the more saline water is found in the vicinity of inlets, while the less saline water is found on the landward sides.

Biotic Components of Estuaries

Estuarine organisms have a wide tolerance for temperature fluctuations, due to the seasonal temperature variations. These organisms are considered to be **eurythermal** (*eury* broad, wide + *therm*). Truly estuarine organisms must also be tolerant of salinity fluctuations. These organisms are termed **euryhaline.** Oceanic organisms, on the other hand, are exposed to relatively constant temperatures and salinities. These organisms are both **stenothermal** (*steno* narrow + *therm*) and **stenohaline.**

The oyster drill preys extensively on oysters. When feeding, scores of oyster drills will invade an oyster bed, climb atop the oysters and bore small pin holes through their shells to suck out the tissue.

In reality since an estuary is a transition zone between the freshwater and the oceanic environments, marine, estuarine, and freshwater organisms are all commonly found in various portions of a typical estuary. In vertically homogeneous estuaries, the distribution of the organisms is governed by their proximity to the oceanic and freshwater sources—the inlets and river mouths—and in relation to the halocline in highly and moderately stratified estuaries. The truly estuarine forms are found in the intermediate portions of the estuary.

Since a number of marine organisms are relatively tolerant to salinity variations, they contribute the greatest number of species to the estuarine community. There are two broad types of marine forms found in estuaries: the stenohaline and the euryhaline organisms. The stenohaline organisms, with a low tolerance to salinity variations, are generally restricted to areas that are immediately adjacent to inlets or to the saltwater wedge, where the salinity is 30 %oo or higher. These stenohaline organisms are generally the same species as are found in the open coastal ocean.

Most of the euryhaline organisms can tolerate salinities as low as 15 %oo and are found considerable distances from inlets. The truly estuarine organisms are also euryhaline and are found in the mid-portions of estuaries in waters whose salinities range from 30 %oo to 5 %oo. These organisms are, however, restricted to estuarine environments and are not found in either fresh or oceanic waters.

Since the range of salinity tolerance for euryhaline organisms is as high as 30 %oo and as low as 5 %oo, it is obvious that many of these species could survive in a highly saline oceanic environment. Since they are absent it is probable that these species are restricted to estuaries by the presence of oceanic predators that cannot tolerate the estuarine conditions, rather than by the higher oceanic salinities. This is true of the oyster, which serves as a food source for the starfish and the oyster drill. Both of these animals prey extensively on the oyster, but cannot tolerate the reduced salinities of estuarine areas. As a consequence oysters are common in, and generally restricted to, estuaries.

Oysters proliferate the less saline portions of estuaries.

The freshwater organisms that are found in estuaries generally cannot tolerate salinities above 5 %oo. These organisms are restricted to the landward portions, where river, stream flow, and surface runoff maintain very low salinities.

The number of species that inhabit estuaries is lower than the number of species to be found in either freshwater or oceanic environments. In terms of freshwater forms, this is due, primarily, to the inability of these organisms to tolerate even small increases in salinity. Since a greater number of species of marine organisms are able to tolerate a reduction in salinity, there is a larger number of marine species found in estuaries. The number is, however, far lower than the number of marine species that are found in truly oceanic environments, since only a relatively few marine species are able to tolerate estuarine conditions.

Summary

Density currents involve a large percentage of the sea's volume. Since these currents are the result of temperature and/or salinity differentials that develop between adjacent water masses, they are often termed thermohaline currents.

The thermohaline circulation travels through the subsurface waters of all of the world's oceans. These currents carry oxygen to the dysphotic zone, where, because of the absence of light, photosynthesis cannot occur. They also remove large amounts of nutrients from the euphotic zone and transport them for great distances throughout the deep ocean.

The current patterns in estuaries are also generally formed in response to density differentials. Depending upon the amount of marine and fresh water that enters these systems, estuaries may be highly or moderately stratified, vertically homogeneous, or hypersaline. The resultant salinity variations in these areas has a direct influence on the biotic distributions in estuaries.

Review

1. What would occur if cold, highly saline water came into contact with warm, less saline water? Explain fully.
2. What is a thermohaline current? Do these currents always move in response to density differentials? Explain fully.
3. Why are the temperatures and densities of the surface water masses important to the subsurface circulation of the ocean?
4. What are the major surface water masses that are involved in the thermohaline circulation?
5. Why is there an absence of Arctic source water in the sea?
6. Sketch the thermohaline circulation. Label all water masses and explain all mechanisms involved in setting these waters in motion.
7. What is the lesser circulation?
8. What is an estuary?
9. Estuaries are classified as highly or moderately stratified, vertically homogeneous, and hypersaline. Explain why.
10. What are eury- and stenohaline and eury- and stenothermal organisms? Where is each type likely to be found in an estuary? Why?

References

Pickard, G. L. (1963). *Descriptive Physical Oceanography.* New York: Pergamon.

Von Arx, W. S. (1962). *An Introduction to Physical Oceanography.* Mass.: Addison-Wesley.

For Further Reading

Duxbury, A. C., & Duxbury, A. (1984). *An Introduction to the World's Ocean.* Mass.: Addison-Wesley.

Gordienko, P. A. (1961, May). The arctic ocean. *Scientific American.*

Kopt, V. G. (1962, September). The antarctic ocean. *Scientific American.*

Stommel, H. (1958, July). The circulation of the abyss. *Scientific American.*

12
Coastlines and Coastal Processes

Key Terms

barrier island
bay-mouth bar
chenier
consolidated coastline
emergent coastline
high-tide horizon
indented coastline
inlet migration
intertidal zone
irregular coastline
low-tide horizon
mid-tide horizon
overwash fan

pocket beach
primary bay-mouth bar
regular coastline
sand spit
sea island
secondary bay-mouth bar
stable coastline
submergent coastline
subtidal zone
Type I coast
Type II coast
Type III coast
unconsolidated coastline

horelines are the point of contact between the marine and terrestrial environments and, as such, are flooded at high tide and exposed at low tide. They are only a part of the coastline or coastal zone, which often extends for miles inland and for hundreds of miles along a shoreline.

Shorelines are dynamic and constantly changing. Their sediments are moved about and form distinctive features, such as sandbars, barrier islands, and sand spits. Since these features often provide protection for the mainland and calm water anchorage for boats, and also frequently create navigational hazards, they are of importance. The coastal areas of the United States are often highly developed and/or valuable recreational areas; therefore, the processes that occur along and shape these areas affect man and his structures.

Coastlines

As noted in chapter 5, coastlines and their associated shorelines are composed of either consolidated or unconsolidated sediments. Unlike the **unconsolidated coastlines,** most of the **consolidated coasts** have been exposed to glacial activity, during which their surface materials were scoured from the land's surface and carried southward. Ultimately these materials were deposited as glacial deposits, such as moraines, when the glaciers halted and began to recede. The coastlines of Oregon and California, however, are consolidated but have not been glaciated. The materials that compose consolidated coastlines must generally be reduced to finer sediments by chemical and mechanical weathering before they are small enough to be moved about by the waves and currents.

Coastlines are also subject to isostatic adjustments and tectonic movements of the earth's crust (chapter 2). Since both of these processes result in the raising or lowering of the land, coastal areas are often classified as **emergent** or **submergent,** on the basis of these movements. For example, the Alaskan coast is considered to be emergent, and many harbors that were navigable in the 1800s are now unnavigable. This situation is the direct result of the isostatic rising of the coastline in response to the removal of weight as the ice cap melted.

Newly formed moraine in the Canadian Rockies. Moraines are mounds of unconsolidated sediment deposited by receding glaciers. In this photo, the glacier has traveled down the mountain as two lobes. As temperatures become warmer the lobes begin to melt and the glacier recedes. Meltwater, the darker areas at the base of the lobes, forms glacial streams. The moraine extends between the meltwater and the small lake in the foreground and marks the original position of the glacier prior to its recession.

The coastline of the northeastern United States is submergent, since a rising sea level has inundated many areas that were previously above water. The Hudson Valley is an excellent example of this "drowned" topography. The present coastline is very irregular, since the topography, prior to the flooding, consisted of hills and valleys.

The remainder of the east coast has a long, broad coastal plain (chapter 2) that extends from the edge of the continental shelf inland. This region has never been glaciated and is composed of unconsolidated sediments. It is considered to be a **stable coastline,** since its position in relation to the ocean has not changed for thousands of years.

Major Coastal Features

Regardless of whether a coastline is emergent or submergent, consolidated or unconsolidated, all coastal areas may be classified as being either **regular** or **irregular** in shape.

Irregular or **indented coasts** are commonly found on consolidated shores and often consist of bold, rocky headlands that project into the sea. A beach, if present, would be located between the headlands and would consist of a thin layer of sand over the bedrock. Such beaches are often called **pocket beaches.** The sediment that composes them is generally the result of the mechanical weathering of the cliff face. The coastlines of Maine and Nova Scotia provide many excellent examples of irregular coastlines that are composed of consolidated sediments. The Oregon–California coastline is also irregular. In this case, however, the headlands consist of volcanic or sedimentary materials and the pocket beaches are often backed by sheer, rocky cliffs. The coastlines of Maine, Nova Scotia, California, and Oregon have only recently been exposed to the action of the sea and are therefore considered to be young, or **Type I coastlines.**

As discussed in chapter 4, the offshore topography focuses wave energy on headlands, and these waves generally strike the shore at an angle. The energy expended by the breaking waves reduces the

A rocky headland composed of consolidated sediment

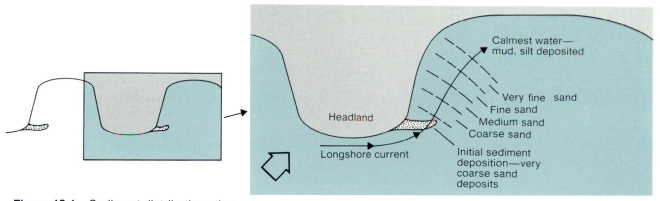

Figure 12.1 Sediment distributions that are commonly found in an embayment with developing sandbars.

boulders and rocks of a consolidated shoreline to finer pebbles and sand that can be moved about by the water in the surf zone. Since they are already small, the sediments that compose unconsolidated coastlines need not be broken down before they can be moved about and transported. Long Island, New York is a terminal moraine that was formed by the deposition of glacially derived sediments. The north shore of the island, though indented, consists of unconsolidated sediments and provides an excellent example of an indented, unconsolidated coastline.

As the headlands erode and cut back, the water accumulates in the surf zone, becomes organized, and forms a longshore current (chapter 5), that begins to move the sediments along the coastline. When this sediment-laden current disperses into the indentation at the tip of the headland (fig. 12.1), the coarser sediments are deposited to form a sandbar that begins to extend across the embayment. The finer sediment is carried farther into the embayment and

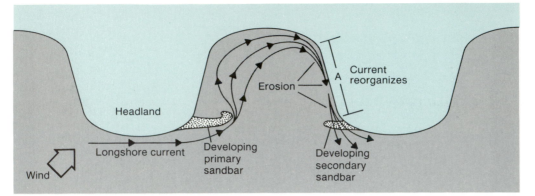

Figure 12.2 The longshore current continually carries water into the indented area. As a result the current reorganizes along the opposite shoreline (A), moves seaward, and picks up and carries sediment from that area.

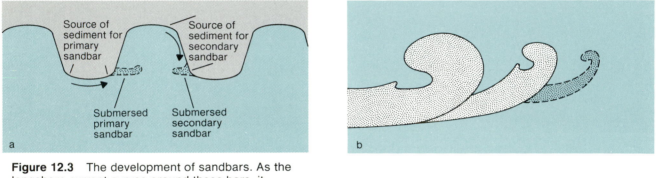

Figure 12.3 The development of sandbars. As the longshore current curves around these bars, it shapes them (b).

is deposited as the current continues to spread out and disperse. Eventually all but the finest sands are deposited on the bottom of the embayment. These fine sands are deposited over the original sands of the pocket beach. A core sample (chapter 3) taken from such a beach would show the recently deposited finer sands in the upper portions and progressively coarser sediment in the lower portions of the core tube-liner.

As the sandbars at the mouths of these embayments continue to build, they slow the longshore current more effectively. This causes a greater amount of sediment to be deposited in these areas, and the sandbars begin to build at a more rapid rate. The subsurface sandbars that grow in the direction

of the longshore current may be considered primary sandbars. As the primary sandbar continues to build, ever-calmer conditions in the embayment follow. This accelerates the deposition of sediments in these calm waters and causes the embayment to shoal.

The wind continues to force additional water into the embayment, causing this water to form a reverse current that moves seaward along the opposite shoreline (fig. 12.2). This current erodes the shoreline and removes fine sand. When the current rounds the opposite headland, it disperses and deposits the sediment as a sandbar (fig. 12.3), which may be termed the secondary sandbar.

Eventually these sandbars build to such a height that they become permanent, abovewater features, which are known as **bay-mouth bars.** The bay-mouth bar that is derived from the primary sandbar is called the **primary bay-mouth bar;** that ex-

Bay-mouth bars

tending from the opposite headland is known as the **secondary bay-mouth bar.** Generally the primary bay-mouth bar is higher, wider, and composed of coarser sediments. The shape and sedimentary characteristics of primary and secondary bay-mouth bars are merely a reflection of the strength of the different currents that formed the bars. The stronger longshore current will carry and subsequently deposit coarser sediment to form the primary bay-mouth bar. The weaker current, moving seaward, is capable of carrying only finer sands, which are then deposited on the secondary bay-mouth bar. Secondary sandbars and bay-mouth bars can also be formed along coastlines when the winds shift and drive the longshore current in the opposite direction for a part of the year, an occurrence that is quite common and is generally seasonal. The current often

reverses in response to gentler summer winds. In these cases the secondary bay-mouth bar is still composed of finer sands and is smaller, narrower, and lower.

The longshore current shapes, as well as builds, bay-mouth bars. As the current moves into an embayment, it disperses at the tip of the bay-mouth bar. This results in a reduction in the current's carrying capacity, and sediment is deposited. It is deposition of this sort that actually causes a bay-mouth bar to extend across the mouth of an embayment. As the current curves around the tip of the bay-mouth bar, it shapes the newly deposited sediment and causes it to form a hook in the same general direction as the dispersing current. Thus, bay-mouth bars extend

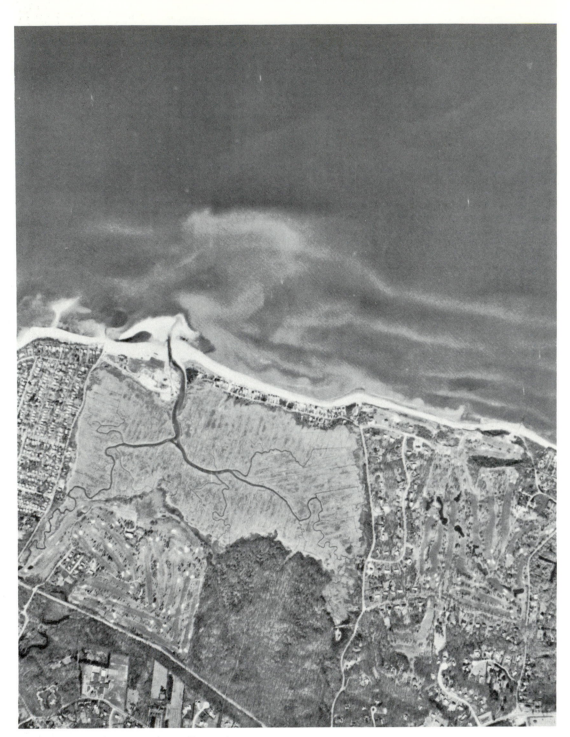

Type III coastline. Note the extensive salt marsh
behind the bay-mouth bars.

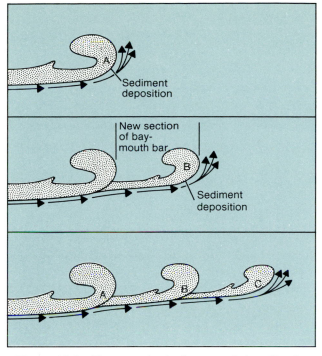

Figure 12.4 The growth of a bay-mouth bar. Sand deposited at point A causes the bay-mouth bar to extend to the right. Water is then diverted in at point B and deposits sediment, which causes the bay-mouth bar to extend farther to the right, and so forth.

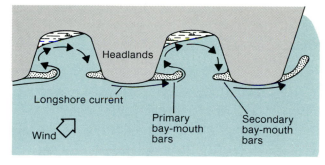

Figure 12.5 An indented coastline with associated bay-mouth bars

across the mouth of an embayment as a series of hooks (fig. 12.4). Once bay-mouth bars emerge, the coastline is considered to be in its mid-age—a **Type II coastline.**

The presence and continued growth of the bay-mouth bars creates increasingly calmer conditions within the embayment. In response finer sands are deposited, causing the bottom to shoal severely. Only the finest sediments—the muds and silts—are carried to the rear, and mud flats begin to develop in these areas as the original pocket beach is ultimately covered by the very fine muds and silts. Once sufficient sediment accumulates, intertidal vegetation, capable of withstanding periodic flooding by marine water and exposure to the atmosphere, will begin to colonize the mud flat.

The longshore current will continue to deposit sand on the bay-mouth bars, and they will continue to extend across the embayment. Eventually they may completely seal this area off. Often, however, a tidal creek will remain to meander through the mud flat, which is densely vegetated by this time. The tidal

creek will provide the sole connection with the marine environment at this time, although storms may wash across the bay-mouth bars periodically. The source of the sediment that builds up the bars is the headland that is located upcurrent of the primary bay-mouth bar. This headland obviously recedes in the process, and the coastline straightens. At this stage, the coast has entered its old age and is termed a **Type III coastline.**

The recession of the headland causes a concomitant recession of the bay-mouth bar. As a result the muds and associated vegetation that have developed behind the initial bay-mouth bars are innundated by seawater as the headland recedes and the shoreline straightens. It is not uncommon to find deposits of silt and mud held in place by the roots of the original vegetation considerable distances seaward of the present locations of the bay-mouth bars and headlands. These deposits mark the location of the original muds and silts that developed in the protected areas behind the first bay-mouth bars.

Should an irregular coastline consist of a series of embayments, this process will be repeated along the entire coast. Each series of headlands will provide the sediment for the primary bay-mouth bars that will form and extend across the mouth of each embayment (fig. 12.5). These headlands will recede, and the coastline will straighten. The erosion will halt only when the wave energy is no longer focused on the headlands; that is, when the coastline is straightened.

Eventually the longshore current will reach land's end, the water will deepen, and the current will disperse and lose velocity. At this point sediment will be deposited and will eventually form a permanent abovewater landform called a **sand spit,**

This sand spit was breached during a hurricane in 1938 (top right). When this occurred the beach to the left became a barrier island. Sand carried by the longshore current filled in the breach and reattached the barrier island.

or simply a spit. Sand spits form in an identical manner and by the same mechanisms as bay-mouth bars. The only difference is that the spit is not growing toward another land mass but is, rather, extending out into the open water as a peninsula. In addition, and in response to the stronger currents, spits will generally extend from the mainland at a sharper angle than do bay-mouth bars. Sandy Hook, New Jersey, and Rockaway, New York, are classic examples of sand spits.

During storms waves will frequently break over spits and bay-mouth bars and may detach them from the mainland. Should these inlets persist, the detached spit or bay-mouth bar would then be called a **barrier island** (fig. 12.6). Barrier islands are often misnamed barrier beaches. In reality bay-mouth bars, sand spits, and barrier islands are all examples of barrier beaches, since they provide protection by presenting a barrier that separates and protects portions of the coastline from the full force of the sea.

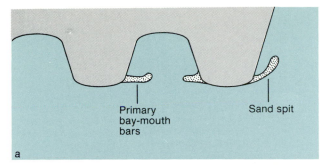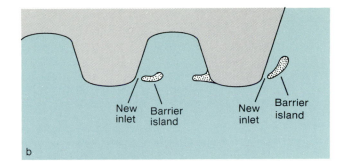

a

b

Figure 12.6 When a bay-mouth bar or sand spit (a) is detached, a barrier island is formed (b).

Primary bay-mouth bars

Sand spit

New inlet

Barrier island

New inlet

Barrier island

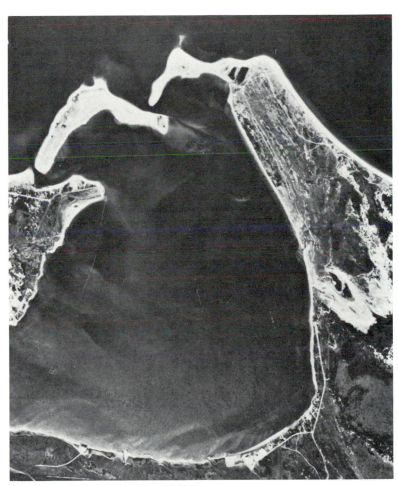

A barrier island (left) and bay-mouth bar (right) provide protection for this harbor.

Barrier islands are most commonly found offshore unconsolidated regular coastlines, although they do occur in limited numbers off the consolidated coasts of New England. In fact the United States has the greatest number of and most well-developed barrier islands in the world. These islands are the major coastal feature from New York to the Texas Gulf Coast.

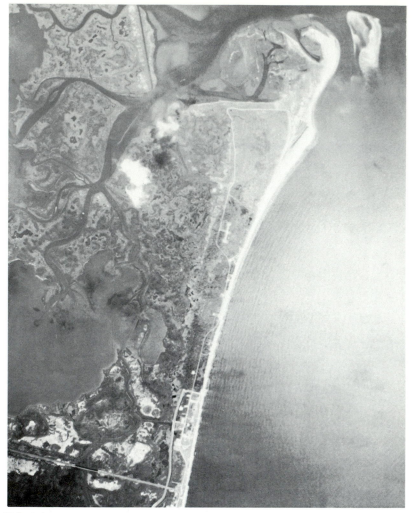

The barrier islands of Virginia

Barrier islands are formed by four major mechanisms. The first, as noted, involves the formation and subsequent breaching of a sand spit. The barrier islands offshore Long Island, Connecticut, and Massachusetts have been formed in this manner.

They may also be formed by the accumulation of offshore sediment on a submarine bar. The continued accumulation of such sediment would eventually enable the bar to build above sea level to become a barrier island. The five barrier islands—Cat, Ship, Horn, Petit Bois, and Dauphin—that separate the Mississippi Sound from the Gulf of Mexico off the Alabama–Mississippi coast, have such an origin.

The barrier islands offshore the Alabama–Mississippi coast

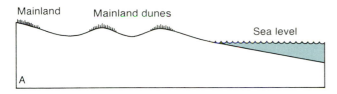

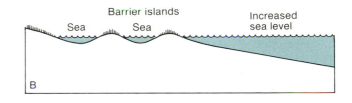

Figure 12.7 A rise in sea level may separate dunes from the mainland and thereby form barrier islands.

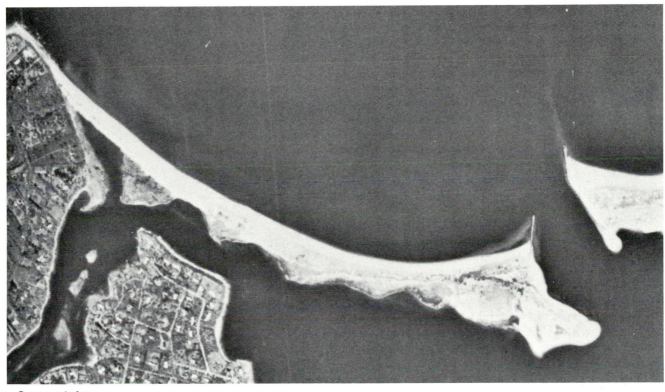

Overwash fans

The third mechanism of barrier island formation involves the flooding of the mainland. When this occurred, preexisting dunes were separated from the mainland when the low areas behind the dunes were flooded (fig. 12.7). As a result lagoons or bays were formed between the newly created barrier island—the dune—and the mainland. It appears that the barrier islands along the southeast Atlantic coast were formed by this mechanism.

River borne terrigenous sediment (chapter 3) may also form barrier islands. This is the origin of the small barrier islands that have been formed on and separated from the Mississippi River delta as a result of the erosion of the delta. In the absence of a delta, barrier islands may also be formed from terrigenous sediment. In these instances the sediments are transported down the coast before being deposited to form a barrier island. Both Popham Beach and Smalls Point off the Maine coast were created in this manner.

All barrier beaches are overwashed during storms. When this occurs the water carries sediment over the beach and deposits it on the bay side, as well as in the bay itself. **Overwash fans** mark areas along

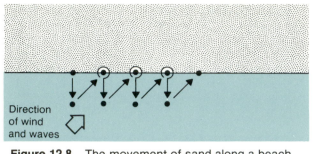

Figure 12.8 The movement of sand along a beach

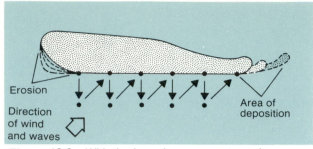
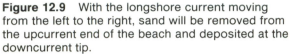

Figure 12.9 With the longshore current moving from the left to the right, sand will be removed from the upcurrent end of the beach and deposited at the downcurrent tip.

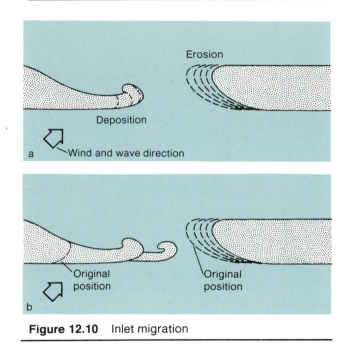

Figure 12.10 Inlet migration

a bay-mouth bar, spit, or barrier island where this has occurred. Along barrier islands that border on the ocean, overwashes may result in the formation of an inlet that intersects and separates the barrier beach. Along the east coast of the United States, hurricanes are a major cause of inlet formation. These inlets are formed from the bay side when large amounts of seawater are driven into the bay by the strong onshore winds. The water is augmented by water brought in as surface runoff from the adjacent landmass during the heavy rains that accompany hurricanes. When the eye of the hurricane passes, the winds shift and drive the bay water toward and onto the barrier beach. This wind-driven water often cuts an inlet through the barrier beach from the bay to the ocean.

Most of the inlets formed during storms or hurricanes are transitory and close shortly after they are formed. Others, however, may provide major pathways by which water enters and leaves the bay. Such inlets will widen, deepen, and persist.

Generally an inlet will not remain stationary but will, rather, migrate along the barrier island. **Inlet migration** occurs in response to the longshore current. As noted in chapter 5, sediment that is carried by this current moves from the beach into the surf zone, back to the beach, and so on, in a zigzag motion (fig. 12.8). Over the long term, there is no net loss or gain of sediment along the beach. There is, however, erosion at the upcurrent tip of the beach and accretion at the downcurrent tip (fig. 12.9). When a barrier beach is intersected by an inlet, these sediments will be deposited on the downcurrent portion of the beach adjacent to the inlet and in the inlet itself. Thus, the inlet interferes with the flow of sediment along the beach.

Since sediment is removed from the longshore current in the vicinity of inlets, the waves that strike the shore on the opposite, downcurrent, side of the inlet do not contain suspended sands. As a result there is a net loss of sediment from the barrier island on the downdrift side of the inlet and an accretion of sand on the updrift side. This will cause the inlet to migrate in the same general direction as the longshore current (fig. 12.10). When inlets migrate in this manner, the channel remains perpendicular to the beach. Inlets of this sort are termed simple inlets.

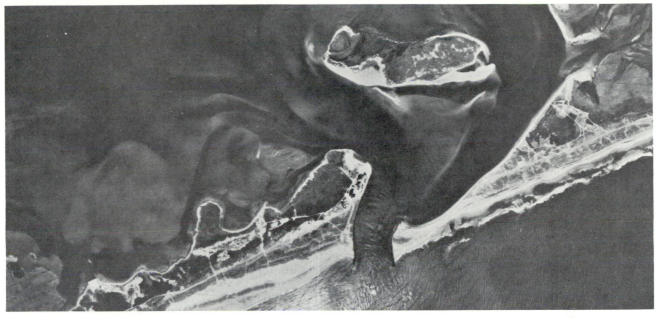

A simple inlet. Note the tidal delta at the mouth of
the inlet on the bay side.

A complex inlet

Occasionally the updrift side of a barrier island that is intersected by an inlet will migrate faster than the downdrift side. In some instances the downdrift side will remain stationary. In either case the velocity of the water that leaves the lagoon on an outgoing tide will force the newly deposited sediment on the updrift side seaward. As a result the updrift side will overlap the downdrift side, and the channel will run parallel to the barrier island. These inlets are termed complex inlets.

The accumulation of sediments in the inlet will cause it to shoal. Newly deposited sediments on the bay side of the barrier island may accumulate to such an extent that they will rise above sea level to form a tidal delta.

As the inlet channel continues to shoal and/or the tidal delta continues to grow, a point will be reached at which the inlet can no longer remove all the sediment from the longshore current. When this occurs some sediments will begin to move past the inlet to the downcurrent portion of the barrier island, and the inlet is said to be bypassing sediment. This process restores the supply of sediment to the beaches downcurrent, which, in turn, will slow or even reverse the erosion along these beaches.

While inlets migrate laterally along barrier islands, the barrier islands themselves migrate toward the mainland. This landward migration generally occurs as a result of major storms.

When overwashes occur and/or new inlets are formed, sand is carried from the abovewater portion of the beach and deposited in the back bay. Since most inlets are transitory and close soon after opening, the calm water bay conditions are restored. The newly deposited sediments are soon colonized by estuarine vegetation such as marsh grass (chapter 13). Simultaneously the barrier island, by trapping wind blown sand from the beach face, soon rebuilds to its original height, only to be cut back when the next overwash occurs. Since this process is repeated continually over long periods of time, sufficient sediments accumulate behind the barrier island to develop into well-defined ridges of submarine sand that run parallel to the barrier island. Eventually this material rises above the sea's surface; once this occurs,

Box 12.2 The Migration of Offshore Bars

Although few measurements have been made on offshore bars, it may be assumed that water breaking on and over the bar moves sediment along the bar, as well as carrying it toward the beach. Should this be the case, sediments will move along the bar in a manner similar to their movement in the surf zone along the beach itself. In other words, the sediments that compose the offshore bar move in response to the longshore current. Hence, a low point in the bar could be considered to be equivalent to a simple inlet and would migrate along the beach in a similar manner.

Since the offshore bar provides considerable protection for a shoreline, coastal erosion could be expected to be most severe along those portions of the beach where the offshore bar is low or absent. Since these inlets through the bar migrate down the beach, the points of erosion along the shoreline can be expected to shift continually in response to the migration of the bar.

sediments are accumulated more effectively. As these processes are repeated, the barrier island will eventually occupy the position of the original submarine ridge and thereby migrate landward.

All available evidence indicates that the barrier islands along the mid- and southeast Atlantic coastline have been migrating in such a fashion since their formation some 8,000 to 10,000 years ago. These islands were formed on the edge of the continental shelf many miles seaward of their present position and have been moving landward since that time, presumably in response to the rising sea level.

It is important to note that these islands, contrary to popular opinion, are indeed migrating rather than eroding, since they move as a geological and biological unit. Headlands, on the other hand, erode from a fixed position.

Sea islands and associated salt marshes of Georgia

Barrier islands often have an associated offshore submarine bar. These bars provide protection to the beach by reducing the velocity of incoming waves. When waves come ashore with high frequency, water is often trapped between the beach face and the offshore bar. This water will eventually find its way back offshore through a low point in the bar and will form a rip current (chapter 5). Such currents often attain velocities of up to 5 kilometers (3 miles) per hour.

Sea islands such as Hilton Head, South Carolina, and Cumberland Island, Georgia, are formed differently from barrier islands. These islands actually consist of two geologically distinct features— a central core that, during periods of lower sea level, was a part of the mainland and an outer, unconsolidated barrier. The barrier may either be attached to the central core with no intervening water or separated by creeks, mud flats, or marsh lands.

The sediments of the central core are approximately 100,000 years old, having been formed during the Pleistocene period, while those of the outer barrier are much more recent and have attached only since the last ice age. The core is very stable, quite high, and is covered by a continental sediment that often supports a mature forest. Only the unconsolidated sediments of the outer barrier move about in response to the longshore current, waves, and tides.

Cheniers, from the French word meaning "oak," consist of ridges of sand and silt that have buried, and hence overlie, salt marshes. Many of the larger, high ridges are covered with dense stands of oak trees, giving rise to the name chenier. The lower portions, between the ridges, are often innundated with water and colonized by estuarine vegetation. Cheniers are common along the Texas–Louisiana coastline and are formed from sediments carried into the Mississippi Sound and the Gulf of Mexico by the Mississippi River.

The barrier beaches of the Texas Gulf Coast

Sea caves

Other Coastal Features

Coastlines also have other features that are interesting, but of minor importance. Many of these features are the result of a differential rate of erosion of a coastline. For example, cavities called sea caves are frequently gouged into the cliff face of a consolidated shoreline. These cavities are formed by erosion along zones of weakness in the cliff's bedrock. Particularly good examples of sea caves are found along the coastlines of northern Oregon and on the Atlantic side of Nova Scotia.

Wave-cut terraces are also found on the faces of these sea cliffs. The terraces consist of horizontal platforms of consolidated rock that are formed below the surf zone in response to the erosion of the cliff

Sea stack

Sea arch

face. Wave-cut terraces are also common above the surf zone and indicate past sea levels and/or iso-static adjustments of the coastline.

Stacks are the more durable remnants of head-lands or cliffs. As the less durable materials erode, these materials remain in the surf zone or farther off-shore to mark the original position of the coastline.

Hopewell Provincial Park in New Brunswick, as well as the coastlines of northern California and Oregon, provide good examples of stacks.

Sea arches are bridges that remain above the eroded central portion of stacks. The sea arches off-shore Cabo San Lucas at the tip of the Baja Peninsula are examples of these structures.

A series of islands connected to the mainland. Note the tombolo on the upper left.

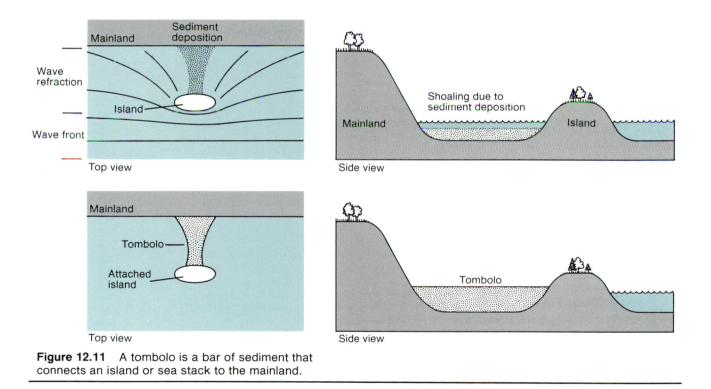

Figure 12.11 A tombolo is a bar of sediment that connects an island or sea stack to the mainland.

Occasionally, islands may be attached to the mainland by sand deposited by coastal currents. The sediment that links the island to the mainland is termed a tombolo (fig. 12.11). The Port-Au-Port Peninsula on the west coast of Newfoundland is connected to the mainland by a large tombolo.

Estuaries

Estuaries, which are integral parts of coastlines, were defined in chapter 11 as semi-enclosed bodies of water that have free connections with the sea. They were classified in that chapter on the basis of their

A typical estuarine fjord

salinity regimes. In order to place estuaries into perspective in terms of coastlines, they may be further classified, on the basis of their geomorphology, into five major types: drowned river valleys, estuarine fjords, tectonic estuaries, bar-built estuaries, and river-delta estuaries.

Drowned river valley estuaries are formed when seawater floods terrestrial areas as a result of either a rise in sea level or subsidence of the coastline. They may be formed along coastlines that consist of low, wide coastal plains, as well as along those that are hilly or mountainous. Chesapeake Bay is an example of the former; the Hudson River Valley, of the latter.

Estuarine fjords are deep U-shaped indentations formed by glacial activity and hence are common on consolidated coastlines. Generally, narrow sills composed of glacial debris are found at the mouths of these fjords. The estuaries of British Columbia, Alaska, and Norway are examples of these systems.

Tectonic estuaries are formed by movements of the earth's crust, which are often accompanied by a large influx of fresh water into the newly formed estuary. San Francisco Bay is an example of such an estuary.

Box 12.3 Estuaries and Evolution

Estuaries are transition zones between freshwater and oceanic environments. Although fresh, estuarine, and oceanic organisms are all found in estuaries, the estuarine forms are euryhaline and eurythermal and could easily survive in the waters of the coastal ocean. They are restricted to the less saline portions of estuaries by the absence of stenohaline predators in these areas.

Truly estuarine forms, able to survive only in estuarine waters, do not, for the most part, exist. This could be due to the fact that estuaries are relatively new environments on earth. If such is the case, truly estuarine animals and plants have yet to evolve.

A tectonic estuary

Bar-built estuaries are actually misnamed, since they may be formed by the building of bay-mouth bars, sand spits or barrier islands, sea islands, or cheniers. They might more properly be named bar-built estuaries when bay-mouth bars separate the open sea from the mainland, spit-built estuaries when spits are involved, and so forth.

These estuaries are typically shallow and frequently receive large amounts of fresh water from river flow. The Great South Bay, behind the barrier island off the south shore of Long Island, and the sounds behind the outer banks of North Carolina, which are actually estuarine areas, are examples of these estuaries.

River-delta estuaries are formed by the deposition of silt deposits at the mouths of large river systems. The delta at the mouth of the Mississippi River and of the Frazer River in British Columbia are river-delta estuaries.

Shorelines

Regardless of the type of coastline, the associated shorelines are profoundly influenced by the rise and fall of the tides. As discussed different portions of the shoreline are inundated daily, depending upon the exact status of the tidal cycle at that particular time. This, in turn, influences the organisms that inhabit this area. As a result different portions of the shoreline have very different assemblages of animals and plants.

The organisms that inhabit that portion of the shore covered only at high tide are subjected to very different conditions than those that are exposed to the atmosphere only during low tide. Furthermore those located below the low-tide mark are permanently covered by seawater and must withstand very

Box 12.4   Groins and Beach Erosion

The barrier beach off the south shore of Long Island, New York, provides an excellent example of the mismanagement of a barrier island and the problems that arise when groins are improperly sited and installed.

This barrier island was extensively developed during the postwar building boom of the 1950s. During this time summer homes, hotels, and other structures were built—some even on the dune itself. This construction, combined with excessive human traffic on and over the dune, caused extensive destruction to the dune.

The loss of the dune was highlighted when severe spring storms repeatedly struck the beach during the 1950s and 1960s. These storms caused severe erosion of the beach face and the berm in front of these structures. The erosion led to the development of a beach and dune restoration plan by the United States Army Corps of Engineers.

The restoration plan called for the widening of the beach and the rebuilding of the dune. Recognizing that the longshore current in this area moved from east to west, the plan specifically stated that if it proved necessary to build groins, they should be built at the far western end of the barrier island first and then be gradually extended eastward. In order to prevent erosion in the groin shadow to the west of each groin, the plan specified that, as each groin was constructed, sufficient sand be placed on its eastern side to assure the continued westward flow of sediment. Placement of sediment in this manner is termed backfilling.

The plan was not immediately implemented. Then when a particularly severe storm struck the midsection of the barrier island in the mid–1960s and many homes were lost, the plan was implemented—but in a haphazard fashion. Contrary to the recommendations, groins were first constructed along the midsection of the barrier beach. To make matters worse, none of the groins were backfilled. As a result severe erosion occurred to the west. Within months the erosion was so severe, even in the absence of storms, that four additional groins had to be built—none of them, however, were backfilled. This merely displaced the erosion farther to the west. It was then concluded that groins do not work, and no further groins were or have been built.

Presently the western portion of the barrier island is very unstable and breaches yearly. In addition many of the homes are actually in the surf zone, with the result that extensive damage occurs with even moderate storms and even during spring tides.

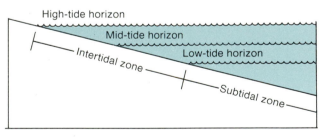

Figure 12.12   The intertidal zone

different conditions than the previously mentioned assemblages. As a result these various regions of the shoreline often differ markedly in terms of the plant and animal populations that are present.

It is customary to divide such areas into the **intertidal** and **subtidal zones.** The intertidal zone is that area of the shoreline that is flooded at high tide and exposed at low tide. Since conditions at the low-tide mark differ greatly from those at the mid- and high-tide marks, it is customary to subdivide the intertidal zone into the **high-, mid-,** and **low-tide horizons** (fig. 12.12). The subtidal zone, though not truly a portion of the shoreline, is the area immediately below the low-tide horizon. This classification is generally used by marine biologists and is discussed in the next chapter.

Jetties are commonly constructed at the mouths of inlets.

Shoreline Stabilization

Although there is little long-term gain or loss of sediment along a barrier beach, there is often temporary, localized erosion or accretion. This occurs when storms strike a coastline and focus their energies along a particular segment of the beach. Since waves strike the beach with increased force and frequency, this area will be cut back temporarily. After the storm passes, the normal sediment flow will be reestablished, and the longshore current will disperse into the newly formed indentation. The longshore current will lose velocity, causing sand to be deposited and eventually rebuilding the beach.

Storm-cut inlets are also generally transitory. Once formed they tend to interrupt the longshore current, shoal, begin to bypass sediment, and eventually reclose. While inlets are open, however, and before they begin to bypass sediment, they may interfere with the longshore current to such an extent that the beaches downcurrent will be deprived of sediment and begin to erode.

Problems arise when man attempts to keep inlets open and/or to hasten the normal rebuilding of the beach. Such attempts often involve the construction of jetties and groins, structures that frequently interfere with the longshore current and exacerbate the erosion.

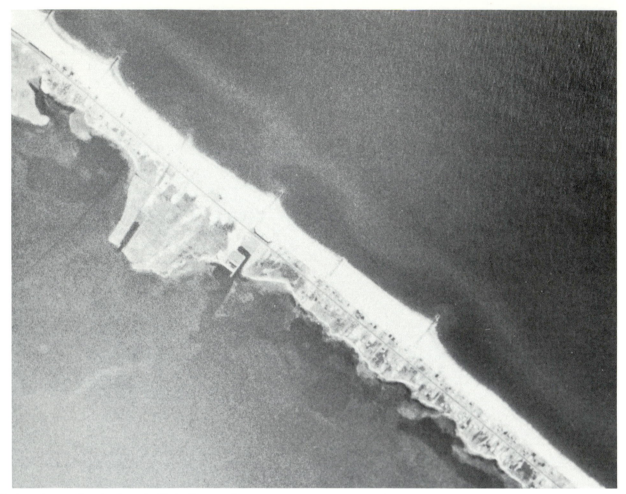

Groins along a barrier beach. Note the erosion
between each groin and to the right of the last
groin. In this case the longshore current is moving
from left to right.

Jetties are merely boulders that are placed at the
mouth of an inlet and extend from the beach into the
offshore waters. Groins are similar structures that are
placed perpendicular to the beach and extend into,
and occasionally beyond, the surf zone. The purpose
of both of these structures is to interrupt the longshore current and cause sediment to deposit on the
upcurrent side of the groin or jetty (fig. 12.13).

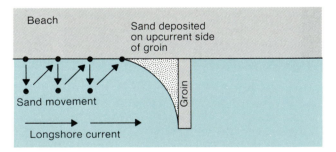

Figure 12.13 Jetties and groins interrupt the
longshore current. Sediment is deposited on the
upcurrent side when the current strikes the
structure and slows.

Beach erosion due to the improper construction of groins.

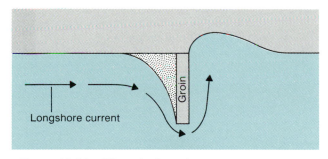

Figure 12.14 When the longshore current reforms, it will cause erosion since all of the sediment was deposited on the upcurrent side.

Although jetties are designed to prevent an inlet from shoaling, they may reduce the longshore transport of sediment to such an extent that the beaches downcurrent of the inlet experience chronic, long-term erosion. Unless properly placed, groins may also cause severe erosion to the beaches immediately downcurrent in the groin shadow (fig. 12.14). Frequently this erosion may extend for thousands of feet downbeach.

Breakwaters are essentially groins that are built offshore and parallel to a beach. These structures are generally built in an attempt to provide calm water anchorage for boats. Since they are designed to reduce the energy of the incoming waves, they also reduce the velocity of the longshore current in the surf zone immediately behind the breakwater. As a result sediment is deposited in this calm water, rather than traveling farther downbeach. Since the wave energy has not been reduced downbeach of the breakwater, the longshore current removes a greater amount of sand from this area than is being deposited. As a consequence the downcurrent beaches may experience erosion.

Summary

Coastal sediments are continually attacked and reworked by the sea. Waves move these sediments into the surf zone, where they are picked up and transported by the longshore current. These sediments are sorted by this current and eventually deposited to form such major coastal features as barrier islands, sand spits, etc. Inlets form, migrate, close, and reform along barrier beaches. A combination of inlet formation, closure, and the overwash of a barrier beach during storms causes the barrier island to migrate landward.

Estuaries—also major coastal features—are formed in a variety of ways. They may form behind a barrier island, sand spit, sea island, or other such feature or be formed by tectonic movements or isostatic adjustments of the earth's crust. Others, the estuarine fjords, are formed directly by glacial activity, while river-delta estuaries, as the name implies, develop behind deltas. Estuaries are called by a variety of different names, bays, sounds, and lagoons being the most common. Regardless of the name, they are all semi-enclosed bodies of water with one or more free connections with the sea.

The construction of groins, jetties, and breakwaters is often harmful and tends to cause or increase erosion. As a result these structures often create much larger problems than they were designed to solve.

All coastlines are extremely dynamic areas that are interrelated by the longshore current. Thus, these areas must be treated as units, since what occurs along one portion will always be reflected by beach conditions downcurrent.

Review

1. Explain how a Type I indented coastline would "age" to a Type III coastline.
2. What is a chenier? How is it formed?
3. What is the significance of overwash fans?
4. How does an inlet migrate?
5. Distinguish between primary and secondary bay-mouth bars.
6. What factors are involved in the formation of submergent and emergent coastlines? Explain fully.
7. What are the major mechanisms involved in barrier island formation?
8. Distinguish between simple and complex inlets.
9. How does a barrier island migrate landward?
10. Discuss the geomorphologic classification of estuaries.

References

Bird, E. C. F. (1969). *Coasts*. Mass.: M.I.T. Press.
Leatherman, S. P. (1983). *Barrier Island Handbook*. Massachusetts: National Park Service Cooperative Research Unit, University of Massachusetts: Amherst.

For Further Reading

Bascom, Willard, 1980. *Waves and Beaches*. New York: Doubleday.
Kassner, J., & Black, J. A. (1983, April) Inlets and barrier beach dynamics: A case study of Shinnecock Inlet, New York. *Shore and Beach*.
Ricketts, E. F., & Calvin, J. (1968). *Between Pacific Tides*. California: Stanford U. Press.

13
Estuarine and Intertidal Ecology

Key Terms

abiotic barrier
biotic barrier
community
competitive exclusion principle
differential reproductive rate
ecosystem
Gause's principle
habitat
interspecific competition
interstitial water
intraspecific competition

lethal limits
mortality
natality
niche
organism
population
range expansion
species
thermally stable
thermally unstable

All organisms are influenced by and, in turn, affect their immediate environment, which is composed of a living, or biotic, and a non-living, or abiotic, component. When organisms with similar requirements come into contact with one another, they will generally compete for various natural resources such as food and living space. One group of organisms will be successful and will proliferate and become common in that area; another will become rare and perhaps disappear from the area altogether.

The ability of animals and plants to successfully interact among themselves, with other animals and plants, and with their abiotic environment is reflected in the size of their populations, their birth and death rates, and other such factors. The organisms and the distributions of these organisms along all coastlines is the end result of these interactions.

Ecological Principles

Ecology is the branch of biology that studies the relationships of the biotic and abiotic factors (chapters 8, 9) that together comprise the environment. The biotic components that are studied by ecologists are **organisms, populations, communities,** and **ecosystems.**

With the exception of the single-celled plankton and bacteria, an organism is considered to be a single individual of a particular species that is composed of cells, organs, and organ systems that function together as a biological entity. Groups of organisms of the same species that live together in a given area are considered to be a population. For example, a group of starfish that inhabit a specific embayment along the coast of Maine would be a population of that species of starfish. Another group of these organisms inhabiting the intertidal zone behind a bay-mouth bar on Cape Cod, Massachusetts, would constitute another population of the same species.

Periwinkles are commonly found in rock-beach communities.

There will, most likely, be other species inhabiting these areas. Generally, for example, the common periwinkle snail is also found in such regions. All populations living in the same area interact among themselves and with the abiotic environment to form a community. A community may be defined as the various populations of plants and animals that inhabit a given area.

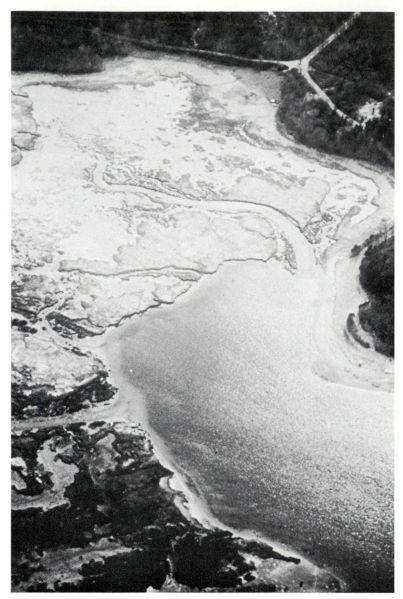

A typical coastal ecosystem consisting of mud flats, salt marshes, tidal creeks, etc.

Several communities located along a specific coastline comprise an ecosystem. For example, the north shore of Long Island, New York, and the southern coastline of Connecticut are awash with embayments, each of which could be considered a community, since each is inhabited by a variety of populations. All these populations interact among themselves and generally with each other, and all are influenced by and influence their abiotic environments. Each of these embayments is interconnected with the offshore waters of Long Island Sound, and these deeper waters, too, contain various populations and communities. All of these biotic and abiotic components interact to form the Long Island Sound ecosystem.

The Long Island Sound ecosystem, in turn, is interconnected with, influences, and is influenced by the mid-Atlantic temperate zone ecosystem. This

An estuarine ecosystem protected by a barrier beach

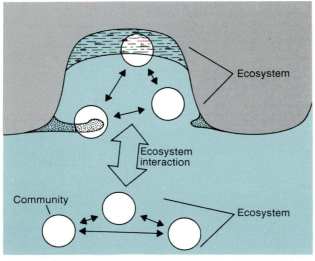

Figure 13.1 Ecosystems are composed of several communities.

oceanic ecosystem is sufficiently different in terms of salinity, temperature, coastal substrates, and so on, to distinguish it from both the north and subtropical Atlantic. Ecosystems may be defined as larger geographic areas that are composed of several communities (fig. 13.1).

The **species** is the basic unit used to delineate populations. A species may be considered to be a group of organisms that are actually or potentially capable of breeding and producing viable, fertile offspring. These offspring, too, must be capable of reproducing in order to perpetuate the species. The horse and donkey, two different species, are able to mate and reproduce viable offspring, yet biologists consider them to be two distinct species. This is because the result of their mating—the mule—is infertile and cannot reproduce.

Box 13.1 An Ever-increasing Problem with Starfish

A knowledge of an organism's niche allows for the proper management of a community or ecosystem, whereas implementing management strategies with an inadequate understanding of niche can lead to costly errors. Such an error occurred in the Long Island Sound, an estuary located between Connecticut and Long Island, New York. In the 1960s starfish became a problem, since they were feeding extensively on commercially valuable shellfish beds. In order to control the starfish, the oystermen devised a starfish "mop" to be dragged along the bottom.

Starfish became entangled in the moplike strands and were swept onto a conveyer belt and carried onto a ship. Once aboard, the starfish were chopped into three or four pieces and the pieces were thrown overboard. The oystermen knew the habitat of the starfish—the shallow bay bottom. They also knew something of the starfish's niche—the fact that starfish are major predators of shellfish. However, the baymen did not realize that pieces of starfish could regenerate if they were large enough. Thus, the process of chopping up the starfish actually served to dramatically increase the starfish population in Long Island Sound.

A population of periwinkles of the species *L littorea,* which inhabits a particular shoreline along the Maine coast, is considered to be the same species as the periwinkle that inhabits a shore along the Long Island Sound. Although the Maine population cannot actually breed with the Long Island population, the potential for reproduction exists; if the Maine population were taken to Long Island waters, reproduction could take place. Hence, both populations are considered to be the same species.

Habitat and Niche

It is convenient to discuss populations in terms of their **habitat** and **niche.** Habitat is defined as the area where a particular organism is likely to be found. For example, the habitat of the periwinkle would be the intertidal zone of rocky coastlines. Habitat tells where the organism lives but little else about it.

Niche is a much more valuable concept and may be defined as the functional role of a population within its community. In order to adequately determine niche, all aspects of an organism's life-style must be considered. For example, it is necessary to determine what it feeds on, what feeds on it, its mode of reproduction, the number of young produced, the type of substrate preferred by the organism, the types of substrate that it can tolerate, and so on.

Intra- and Interspecific Relationships

The organisms that comprise a given population compete among themselves for a variety of factors such as food, actual physical living space, and, in some cases, mates. This is said to be **intraspecific competition,** since it occurs between organisms of the same species. Those organisms in the population that are the better competitors will obtain the most food, the preferred living space, etc. These superior organisms are said to be the most fit and therefore they will, in all likelihood, begin to breed earlier with other superior individuals, producing more offspring than the less fit members of the population. In addition there is a high probability that a greater percentage of these offspring will be superior and so, have a higher survival rate than the offspring resulting from the mating of less fit individuals.

These superior offspring will also begin breeding earlier and will breed with other superior individuals to produce superior offspring; and the cycle will be repeated. Over the long term, therefore, the progeny of the superior organisms will have a numerical, as well as a physical and/or physiological advantage and will outcompete the less fit individuals in the population. Essentially this is an example of Charles Darwin's survival of the fittest, which occurs as a result of a **differential reproductive rate** in which the fittest organisms produce the most progeny, the fittest of which will survive to ultimately dominate the population.

Box 13.2 478 Million Eggs Per Sea Hare

The sea hare is a shell-less snail common to the estuaries of California. This animal is hermaphroditic; that is, each sea hare contains both male and female reproductive systems. A given sea hare may perform as the male during one copulation and the female at the next, or the organism may carry out both roles simultaneously. At times groups of sea hares arrange themselves in an arc in which each animal has its penis inserted in the vaginal opening of the animal ahead.

It has been estimated that a single sea hare is able to produce 478 million eggs during a single breeding season. Obviously not all the eggs mature. In all likelihood the vast majority do not survive for a variety of reasons. The few that do are the fittest and will perpetuate the sea-hare population.

Sea hare

Weaker individuals will either die out or, if they are motile, may move to another location where intraspecific competition is less severe. The entire process results in the continual improvement of the population. This process is rather passive; seldom, if ever, do the organisms engage in active, physical combat.

Within a community individuals of different species may compete with each other for food, living space, and in some cases, breeding areas. When this occurs the competition between different species is termed **interspecific competition.** When species compete interspecifically, the less fit population will be outcompeted, its reproductive rate will fall, and the fitter species will proliferate and dominate the community. The individuals that comprise the less fit population will either decrease in number or die out. If motile, the population may move to a different area where there is less competition.

Both intra- and interspecific competition result in the most fit organisms becoming more efficient in obtaining the natural resources that they need to carry out their life cycles. Should the less fit organisms be able to find and move to another area, they may be forced to employ different feeding strategies, adopt an alternate food supply, etc. Thus, the superior organisms may become more specialized in terms of food, while the less fit organisms may become more versatile.

Excessive specialization may actually be to the species' disadvantage. For example, a species of bird, the Everglades kite, has developed a very specialized beak that enables it to feed exclusively on snails that inhabit the Everglades in Florida. As the Everglades have become developed, the snail population has drastically declined, resulting in such a dramatic decline in the kite population that many fear that this species will become extinct. As this example illustrates, when a species becomes modified in response to natural factors, remarkably favorable adaptations may result. Man, however, frequently changes conditions so rapidly that a species cannot respond.

In 1932 an obscure Argentinian biologist, Angel Cabrera, related niche to competition in what is now known as the **competitive exclusion principle.** This principle holds that no two species can occupy the same niche at the same time in the same community without triggering competition. When species begin to compete, one of three things must occur: the poorer competitor may be forced out of the community, it may die, or one or both of the competing species may be able to modify their lifestyles to a sufficient extent to avoid competition. Should the latter occur, each species or the weaker competitor, depending upon the circumstances, will utilize a more restricted niche. For example, competition may force the weaker species to feed only on a single species of prey, while in the absence of competition a variety of different prey species might have been utilized. In most communities niche contraction is generally the end result of interspecific competition. Most likely, only the poorest of competitors die or are forced out of an area. However, the superior competitor will have a higher reproductive rate and will dominate the community.

The competitive exclusion principle is often termed **Gause's principle** after G. F. Gause, who made the correlation independently but at a later date than Cabrera. Gause's work was more widely circulated and became better known. Cabrera, however, not only predated Gause, but also stated the principle more clearly and concisely.

While interspecific competition often leads to a niche contraction within the less fit population, intraspecific competition, by bringing about a higher reproductive rate among the more fit individuals within a population, often causes the population to expand its range.

The genesis of **range expansion** can best be observed when a few individuals move into a previously unoccupied area that contains all the natural resources required by that species. When this occurs the population will, initially, begin to increase at a modest rate as the individuals accustom themselves to their surroundings. Following this period the population growth rate will increase dramatically and then, ultimately, level off (fig. 13.2).

Should the birth rate and the death rate, respectively termed **natality** and **mortality,** remain the same throughout time, it follows that some individuals must be leaving the area. Migration among motile populations generally occurs as a result of

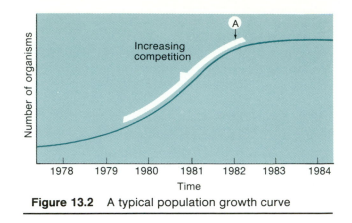

Figure 13.2 A typical population growth curve

intraspecific competition. The young and older animals are generally the poorer competitors and are forced to move into other areas in search of food and other needs. When this occurs the population has expanded its range (fig. 13.3).

Range expansion will continue until halted by either a **biotic** or **abiotic barrier.** For example, the periwinkle snail requires a rocky substrate; its food supply, various species of algae, is found attached to the rocks. Thus, the periwinkle can only expand its range along rocky shorelines, since its food supply is limited to these areas. Should it attempt to colonize a muddy shoreline and utilize another food supply, the periwinkle would come into interspecific competition with the mud snail. This species, as a result of its own continual intraspecific competition, is well adapted to these muddy areas and would, in all likelihood easily outcompete the periwinkle.

In many cases a population undergoing range expansion will enter an area that is suitable but is inhabited by another species with similar niche requirements. Should this occur interspecific competition will arise and the weaker species may be forced to contract either its range or its niche.

It is important to note that niche contraction enables the weaker species to remain in the same area. Range contraction, on the other hand, occurs when an organism is forced from an area by interspecific competition. In general range expansion is the result of intraspecific competition, while interspecific competition leads to either range or niche contraction.

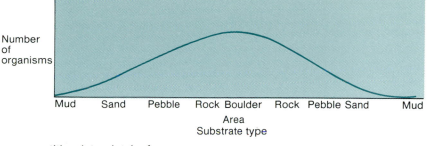

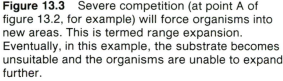

Figure 13.3 Severe competition (at point A of figure 13.2, for example) will force organisms into new areas. This is termed range expansion. Eventually, in this example, the substrate becomes unsuitable and the organisms are unable to expand further.

Biotic barriers to range expansion include an absence of a suitable food supply or the presence of other species that the expanding population may come into competition with. The most common abiotic barriers to range expansion are substrate type, tidal range, temperature, and salinity. Of these, substrate, tidal range, and salinity are important locally, while temperature limits the range of organisms over a broad geographical area. Temperature effects are also important locally in estuarine areas, as well as in the intertidal zone.

Abiotic Factors

Temperature

The large-scale geographical effects of temperature on biological distributions are very pronounced along the east coast of the United States. This is evident in the considerable biological diversity of the area. The temperate Atlantic Ocean between Nova Scotia and Cape Hatteras, North Carolina (fig. 13.4), encompasses only 9° of latitude, yet has biotic components of both the polar and tropical Atlantic. This large variety in life forms is due to the path of the Gulf Stream (chapter 10), which moves as a coastal current along the shores of Florida, Georgia, South Carolina, and southern North Carolina. At Cape Hatteras the Gulf Stream moves offshore and continues northward as an offshore current. As a result the coastal waters to the south of Cape Hatteras are warm and exhibit a temperature range of only 10°C (18°F)

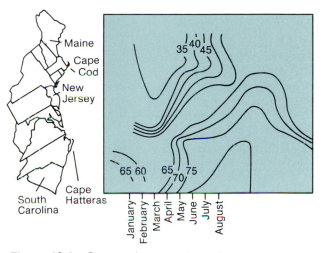

Figure 13.4 Seasonal temperature changes between Maine and South Carolina. Note the penetration of cold water southward in the winter and warm water northward in the summer.

between their summer highs and winter lows. The coastal waters from Cape Hatteras to Cape Cod, Massachusetts, on the other hand, are subject to wide temperature fluctuations and vary as much as 20°C (36°F) annually. The coastal waters to the north of Cape Cod move southward from the colder northern seas. The waters from Nova Scotia to Cape Cod are continually cold but, like their counterparts to the south of Cape Hatteras, exhibit a narrow temperature range of 10°C.

Water masses exhibiting small temperature ranges are said to be **thermally stable;** those having wide temperature variations are **thermally unstable.** Thus, it can be said that the waters from Cape Hatteras to Cape Cod are thermally unstable and are located between two thermally stable water masses, one cold and the other warm.

These different temperature regimes have important biotic implications and cause the waters of Cape Cod and Cape Hatteras to behave as thermal barriers that prevent the permanent range expansion of southern species northward and northern species southward. For example, in the winter the waters from Cape Cod to Cape Hatteras cool to such an extent that northern species, such as cod, are able to migrate southward. Northern plankton are also carried southward by the cold-water currents, but theirs is a passive migration. As the water warms in the late spring, the cod again move northward; the plankton, since they cannot withstand the warmer water temperatures, die.

The reverse occurs in the summer when warm water penetrates to the north of Cape Hatteras. Warm-water species are able to migrate as far north as Cape Cod. The cold-water thermal barrier generally prevents migration farther north, although at times some southern species are found in the Gulf of Maine. When the waters cool in the fall, the fish again move southward, while the plankton, adapted to warm water, perish.

Since the seasonal temperature changes at Cape Hatteras and Cape Cod prevent the permanent range expansion of northern and southern species, these capes are termed biogeographic change points. These biogeographic boundaries are not, however, absolute; rather, they act as selective filters. For example, the Virginian province (chapter 10) has few native species but consists primarily of more northerly and southerly components. This province is said to be impoverished, meaning that there are few permanent northern and southern residents in it. In reality, the Virginian province is a transitional area between the more thermally stable Carolinian province to the south and the more northerly or Boreal region.

Box 13.3 Anchor Ice

Anchor ice may form during very cold winters in the temperate zone. When this occurs, water crystallizes about an object on the bottom. Additional crystals form and may surround and freeze sessile organisms.

Anchor ice caused a high mortality in the blue-claw crab and soft-shell clam population during the severe winter of 1977–1978, when large quantities of anchor ice formed in the Great South Bay, a barrier island-built estuary located off the south shore of Long Island, New York.

On a more localized scale, temperature effects are important abiotic factors in estuarine areas and, on a still smaller scale, in the intertidal zone. Estuaries contain a smaller volume of water than the adjacent, offshore seas. Since smaller volumes of water warm and cool more rapidly in response to atmospheric temperatures, estuarine water temperatures vary seasonally. These temperature ranges are altered by the input of fresh water from rivers, streams, and surface runoff. Since the temperature of these freshwater systems closely approximate the atmospheric temperatures, they carry cold water to the estuary in the winter and warm water in the summer. This reinforces the temperature differences between estuarine waters and the coastal sea. Moreover rivers, streams, etc., enter along the landward fringe of an estuary, while more saline water enters via inlets and such, resulting in a small temperature range in the seaward portions of an estuary and a wide temperature range toward the landward portions. Since the fresh water is less dense than either the seawater or estuarine water, it tends to layer at the surface. Stratification of this type is responsible for the vertical distribution of temperature in some estuaries. All these factors cause estuaries to be warmer in the summer and colder in the winter than are the adjacent coastal waters.

Temperature also plays an important role in the intertidal zone. These areas are periodically covered by water at high tide and are exposed to the atmosphere during low tide. Since air has a much wider daily and annual temperature range than water with its high latent heat (chapter 6), intertidal organisms are exposed to wide temperature variations. Indeed, the temperature range of the air frequently exceeds the **lethal limits** for marine organisms. Should intertidal organisms be exposed at low tide, when the air temperature is at its maximum or minimum, lethal limits may be exceeded and the organisms will be killed. Even if the lethal limits are not reached, the air temperature may be sufficiently low or high to so weaken the organisms that they will become more vulnerable and die of secondary causes, such as a reduced food supply, etc.

Salinity

Salinity variations also have profound effects on the distribution of plants and animals, particularly in estuaries and in the intertidal zone. The classification of estuaries on the basis of salinity patterns and the effect of salinity variations on the distribution of estuarine organisms are discussed in chapter 11. In addition to these relatively large-scale effects, the tidal range, the Coriolis effect, and seasonal rainfall and temperature regimes also influence estuarine salinity patterns.

In estuaries having a large tidal range seawater will be carried far into the estuary on a rising tide, while on a falling tide, subsurface fresh water may penetrate considerable distances seaward. In these cases certain regions of the estuary will have a salinity regime that ranges with the changing tides from highly saline to virtually fresh water. In these areas of maximum salinity fluctuation, the salinity range over a six- to twelve-hour tidal period may exceed the annual salinity range for the entire estuary. The motile animal populations may consist of freshwater species during low tide and marine species during high tide. The sessile organisms, if present, must be euryhaline (chapter 11).

The Coriolis effect is the deflection of moving waters due to the rotation of the earth. It will be recalled that these waters are deflected to their right in the Northern Hemisphere and to their left in the Southern Hemisphere. In the Northern Hemisphere when saline water travels into an estuary, it is deflected to its right; the fresh water, moving seaward, is also deflected to its right—along the opposite

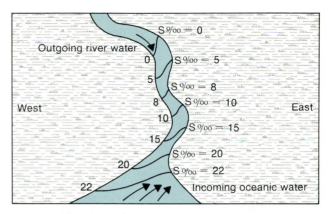

Figure 13.5 The Coriolis effect deflects moving waters to the right in the Northern Hemisphere. As a result river water is deflected to the west side and oceanic water to the east side when it enters rivers flowing southward.

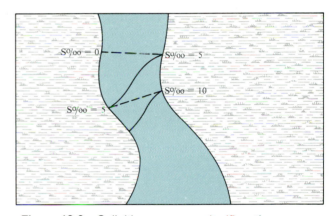

Figure 13.6 Salinities may vary significantly on opposite shorelines due to the deflection of oceanic and river water.

shoreline (fig. 13.5). To observers facing the estuary the seawater would be deflected to their right and the fresh water would be deflected to their left. As a result of the Coriolis effect, there may be two points on the opposite shorelines of an estuary that will have very different salinities (fig. 13.6). These areas may be marked by different animal and plant populations as well. In the Southern Hemisphere the deflection of the saline and fresh water would be

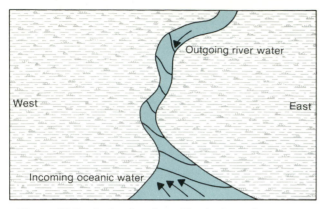

Figure 13.7 In the Southern Hemisphere the waters are deflected to their left.

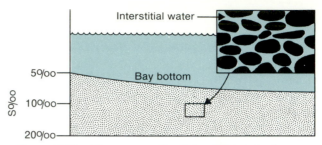

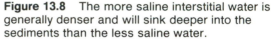

Figure 13.8 The more saline interstitial water is generally denser and will sink deeper into the sediments than the less saline water.

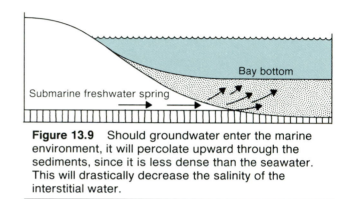

Figure 13.9 Should groundwater enter the marine environment, it will percolate upward through the sediments, since it is less dense than the seawater. This will drastically decrease the salinity of the interstitial water.

reversed (fig. 13.7), as would the points of high and low salinity along the opposite shorelines—the incoming saline water would be deflected to the observer's left.

Seasonal changes in the salinity of an estuary are generally correlated to seasonal changes in temperature and/or river flow. In the temperate zones the warmer summer temperatures lead to a higher evaporation rate and therefore a higher estuarine salinity in the summer. Similarly in areas where the river flow and surface runoff are altered seasonally due to precipitation patterns, the estuarine salinity will increase during dry periods and decrease during periods of greater rainfall.

Salinity variations also affect the distribution of burrowing organisms. The sediments of the subtidal and intertidal zones of unconsolidated coastlines are most frequently composed of mud and/or sand. Water known as **interstitial water** saturates the spaces, or interstices, between these sediments. Similarly, the burrowing organisms that inhabit these interstitial spaces are termed interstitial organisms.

Interstitial water originates from the overlying seawater that permanently covers the subtidal sediments or that periodically inundates the intertidal zone on a rising tide. Since the more saline water is generally denser than the water of lower salinity, the most saline water will be found immediately above the bottom. It is this saline water that percolates into

the sediments and forms the interstitial water. As a result the bottom-dwelling organisms of the subtidal and intertidal zones will be continually exposed to saline water (fig. 13.8).

There are, however, two exceptions: the tidally dominated estuaries and those sub- and intertidal areas that receive significant inputs of groundwater. As discussed previously tidally dominated estuaries exhibit maximum salinity fluctuations, causing wide variations in the salinity of their interstitial water. Along many coastlines groundwater, formed by rainwater percolating into terrestrial sediments, moves laterally and intersects the earth's surface in the intertidal or subtidal zone. When this occurs, the less dense groundwater travels upward through the denser saline interstitial water. The burrowing organisms may then be exposed to a continual flow of rising fresh water. This reduction in the salinity of the interstitial water may be sufficient to eliminate bottom-dwelling organisms in such areas (fig. 13.9).

In the upper portions of an estuary, the intertidal sediments are inundated only during high tide, when more saline coastal waters are carried up into the landward portions on the rising tide. By contrast the more seaward portions of the intertidal zone are covered by water of a lower salinity, due to the increased amount of river water that moves down the estuary on a falling tide. As a result the interstitial salinity is higher in the high-tide horizon than it is in the low-tide horizon, which often permits interstitial marine organisms to penetrate farther up an estuary in the high-tide horizon than in the low-tide horizon.

The salinity of interstitial water also plays an important role in the distribution of intertidal burrowing organisms on a barrier beach. These beaches develop on the seaward side of an estuary, generally far-removed from any freshwater input from rivers. Since they are alternately exposed to the atmosphere and covered by seawater, it might be expected that the organisms inhabiting these beaches would be stenohaline (chapter 11) and able to survive only in relatively saline water. Rather these organisms, particularly those living on or toward the surface of the sediment, are euryhaline and can successfully withstand salinity variations of 15 %oo. Although the organisms are exposed to relatively saline water on a rising tide, they are also subjected to fresh water during periods of rainfall when the tide is low. Since fresh water is less dense than the interstitial seawater, the fresh water will not sink into or below the interstitial seawater. As a result only the sediments near the surface contain interstitial water that is either totally fresh or greatly reduced in salinity. Below the freshwater–seawater interface, the salinity of the interstitial water remains similar to that of the water that overlies the sediments during high tide. Due to these differences in the salinity of the interstitial water, euryhaline organisms are found on the surface and in the sediments nearest the surface of intertidal barrier beaches, while stenohaline organisms are found in the deeper sediments.

Tides

There are three major tidal effects exerted on the animals and plants that inhabit the intertidal zone: duration of the tide, the time of day during which a particular tide occurs, and the rhythmic periodicity of the tides.

The duration of the tide controls the amount of time that intertidal organisms will be exposed to the atmosphere. For example, the animals of the low-tide horizon are subjected to very different exposure times than those of the high-tide horizon. It is during the time period that these organisms are exposed that they are susceptible to desiccation and lethal atmospheric temperatures. The longer the period of exposure, the greater the probability of an organism's desiccating or encountering a lethal high or low temperature. Increased exposure times may also weaken the organism to such an extent that it may die of secondary causes. In addition most of the sessile marine organisms can feed only when immersed in water. As a result the sessile animals of the low-tide horizon have a much greater opportunity to feed, as compared with those of the high-tide horizon. This is commonly reflected in the size distribution of a population that inhabits the entire intertidal zone. The organisms that inhabit the low-tide horizon are generally larger than their counterparts in the high-tide horizon. This phenomenon is commonly observed in barnacle populations; those of the low tide horizon are invariably larger than those of the high-tide horizon.

The time of day during which a particular tide occurs is also important, when considered in conjunction with the possibility of an organism's encountering a lethal temperature. There is a greater probability of a tropical intertidal organism's being subjected to a lethally high atmospheric temperature if low tide occurs at noon rather than at midnight. The reverse would be true in the polar regions, where there would be greater probability of an organism's encountering a lethal low temperature if exposed at midnight rather than at noon. In the cooler temperate regions, lethal temperatures are most likely to occur when the organism is exposed during the day in the summer and at night in the winter. During these seasons, lethal highs and lows, respectively, are most likely to occur. Finally, the predictability of tidal cycles appears to have induced biological rhythms in many marine organisms. As noted many sessile intertidal animals are able to feed only when immersed in water. Moreover,

Grunions spawning

Mud flat and associated marsh grass

many, if not most, of the organisms that inhabit the intertidal zone become quiescent when exposed to the air and resume their normal activities only when their habitat is flooded. Furthermore some animals such as the grunion, a fish that inhabits the coastal waters of the Pacific Ocean, time their spawning to coincide with the summer spring tides (chapter 4).

Substrate

Perhaps no other abiotic factor plays such an obvious role in controlling the distribution of marine organisms as does sediment type. For the purposes of describing habitat, all sediments can be divided into either consolidated or unconsolidated sediment types. Habitats composed of consolidated substrates include glaciated coastlines such as Nova Scotia, where the underlying bedrock has been exposed and meets the sea in the intertidal zone, as well as the rocky shoreline of Maine. In such areas the finer sediments, ranging from coarse sand to silt, are generally carried to the deeper waters and cover the bedrock offshore. Rock jetties, groins, and breakwaters (chapter 12) are examples of artificial, man-made consolidated habitats. Since these structures are generally built to stabilize unconsolidated shorelines, they enable organisms requiring a consolidated substrate to populate shorelines where they would not normally occur.

Unconsolidated substrates are commonly found along unglaciated coastlines. These substrates consist of those materials that are moved about by the normal waves and currents that occur in any given area. Thus, sediments ranging from pebbles to clay and silt comprise unconsolidated substrates. Sediments play such an important role in biotic distributions that communities are often designated by giving their dominant sediment type.

Substrate–Community Interactions

Ecologists frequently describe and define communities by naming them after the most obvious biotic or abiotic factor. Using this method, it is also possible to infer the types of organisms that would be present in a given community. For example, a mud flat behind a bay-mouth bar is often a very obvious feature. This area could be termed a mud-flat community, naming it for its most evident abiotic feature. The name mud-flat community indicates that the sediments are very fine and that organisms such as barnacles, which cement themselves onto the substrate, are not to be expected. Rather, burrowing animals would be the likely inhabitants, along with mud snails.

Similarly, a rocky shoreline is a common and obvious feature along wave-swept coasts with strong currents. In these areas the breaking waves and currents remove the finer sands, clay, and silt. A rocky

Rocky shoreline

The soft-shell clam is a common inhabitant of sand-beach communities.

shoreline indicates an area where large amounts of wave energy are released; these areas are often called rock-beach communities. Rock-beach communities are composed of organisms that are adapted to clinging to the rocks in order to prevent being swept away. Barnacles, starfish, periwinkles, and seaweeds with holdfasts are common populations of a rock-beach community. This community, too, is named for its most prominent feature, the abiotic substrate.

The tip of a bay-mouth bar is an area where the longshore current deposits considerable quantities of coarse sand as it travels into an embayment. Such an area has a high sedimentation rate, and organisms living on the bottom are in danger of being buried. Organisms living in this sand-beach community must be capable of rapidly burrowing through the substrate to keep pace with the rapid sedimentation. Typical populations found in a sand-beach community are soft-shell clams and ribbon worms.

As noted in chapter 12, mud-flat communities may be rapidly invaded by vegetation that is capable of surviving in the intertidal zone. Once a mud flat reaches a certain height, this vegetation moves in and rapidly obscures the mud flat. When this occurs the most obvious feature is the vegetation; in the temperate zone, this would be the salt-marsh grass, while in the tropics it is the mangroves. Hence, in the temperate zone it is common to have salt-marsh communities, while in the tropics mangrove communities develop. These communities are named for their most obvious biotic assemblage, the vegetation.

a

The salt-marsh grass (a) and mangroves (b)

b

A mussel-bed community

In many cases rocky shorelines provide points of attachment for either the blue mussel or the barnacle. Often these animals grow so densely on the rocks that they become the most obvious feature. Such communities would be named blue-mussel or barnacle communities, since these animals are the most obvious features.

Often the success of a population within a given area causes a rapid change in substrate type. For example, when blue mussels move onto a gently sloping rocky substrate, they often do very well and blanket the rocks. The preexisting mussel shells then provide a substrate for other mussels, and soon mussel upon mussel upon mussel will cause the water bearing finer sediments to slow to such an extent that these fine sediments will be deposited over and cover the mussel population. The rocky shoreline, in this case, has been converted into a mud flat, and the community will often be invaded by vegetation and become a salt-marsh community. This is an excellent example of the abiotic environment fostering a particular biotic component, which then converts the area into a different substrate type, which, in turn, brings in a totally different community.

Since the substrate determines the types of organisms that will be present along a given shoreline, it is often convenient to divide these areas on the basis of substrate type. Basically there are only two types of substrate: consolidated and unconsolidated. As noted consolidated shorelines can be composed of bedrock, boulders, or rocks—those materials not moved about by normal currents and waves. Unconsolidated shorelines consist of the various sands, as well as clay and silt. Beaches that are composed of pebbles are a special case, since they provide points of attachment for organisms typical of rocky shores, yet are frequently moved about by normal waves and currents as are sandy and muddy substrates.

Primitive molluscs called chitons are common on rocky tropical coasts.

Rocky Shores

The most obvious feature of rocky shorelines is the zonation of various populations of animals and plants. This is due to the distribution of specific populations in the high-, mid-, and low-tide horizons.

The high-tide horizon is an area of extremely stressful conditions. It is exposed to the atmosphere for long periods of time, causing the temperatures in this zone to fluctuate widely. In addition dissolved plant nutrients are scarce and are available only when the area is inundated during high tide. Due to these strenuous conditions, the high-tide horizon is sparsely populated.

The major producers in this zone are various species of green and blue-green algae and a type of blackish colored lichen. Lichens are associations of a fungus and single-celled algae. The fungus provides a point of attachment for the algae, while the algae, via photosynthesis, produce nutrients for the fungus. In this way both organisms benefit from the association, which is said to be symbiotic. Plankton populations are transitory and are present only when this zone is covered by seawater. The dominant grazers, feeding solely on the algae and lichens, are the periwinkles and limpets. Immediately below the algae is a conspicuous area inhabited by barnacles. Since these animals are filter feeders, they are able to feed only when the tide is in and plankton are present. When the area is exposed to the atmosphere, the barnacles seal themselves within their shells and protect themselves from desiccation.

Barnacles filter plankton from seawater. Note the filtering apparatus extending from the barnacles.

The midtide horizon is covered by water more frequently, the temperatures are less variable and the nutrients more plentiful. As a result this area is more densely populated than the high-tide horizon. Dense growths of macroscopic green, red, and brown algae are present and provide food for the grazers, which consist of populations of mussels, limpets, and chitons.

The dominant animal of the midtide horizon is the mussel, a filter feeder. This animal attaches to the substrate by elastic byssal strands. Since these strands are elastic, the mussel is able to orient itself to a limited extent in the water column, enabling it to feed at different tidal levels. This ability gives the mussel a competitive advantage over other filter feeders, such as the barnacle, which is cemented to the substrate and cannot alter its position. Mussels can, and often do, overgrow and outcompete the more sedentary barnacles. They do not, however, totally eliminate the barnacles, since mussels are extensively preyed upon by starfish, which are also present on rocky substrates. In addition, mussels in the temperate zone often suffer a high mortality during the winter when they are scraped from the substrate by sea ice. The more firmly attached barnacles do not, apparently, suffer such high mortality. Obviously both biotic and abiotic factors serve to limit mussel populations in the midtide horizon.

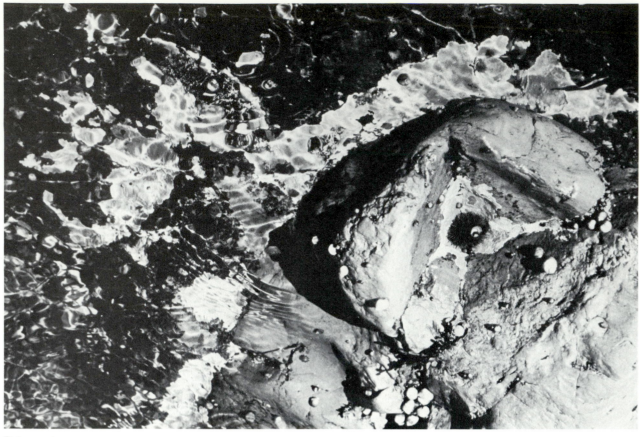

Tide pool

The low-tide horizon is covered most frequently by the tides, with the result that the plant nutrients nitrate and phosphate are most consistently available. In response there are very high populations of the macroscopic green, red, and brown algae in this area. Moreover, since the low-tide horizon is frequently inundated by seawater, plankton are generally available. Thus, food for the grazers is abundant and the numbers of these animals, as well as of the predator populations, are often high. Due to the abundance of food and the stability of the temperature and salinity regimes, neither biotic nor abiotic factors seem to limit the populations of the organisms of the low-tide horizon. Further, interspecific competition, save for living space, is generally insignificant; hence, no single species dominates in the low-tide horizon. As in the midtide horizon, however, predation may be a major factor in limiting the numbers of some species.

In general, biotic factors serve to limit the success of populations in the low-tide horizon, while abiotic factors, primarily salinity and temperature variations, are important in the high-tide horizon. A combination of both biotic and abiotic factors affect the populations of the midtide horizon.

Tide pools are a particular type of habitat that are often found on rocky substrates. Although they may be found at any level in the intertidal zone, those found in the low-tide horizon are flooded very often. As a result their biota are similar, if not identical, to the surrounding plants and animals of the low-tide horizon. The tide pools of the mid- and high-tide horizons, on the other hand are flooded less frequently and so are exposed to the air for longer periods. This exposure leads to a much different abiotic regime and results in the biota being subject to widely fluctuating temperatures, salinities, and oxygen levels.

There is a very small volume of water in a tide pool and when the tide is out, this water warms and cools much more rapidly than the surrounding sea.

In the summer the water in a tide pool may reach a lethally high temperature, while in the winter it may fall to lethally low temperatures—or even freeze completely. When the tide comes in, tide pools are flooded with seawater of a very different temperature—in the summer with cooler water and in the winter with warmer water. In order to survive, the biota of tide pools must be able to withstand both wide temperature variations and rapid temperature changes.

A similar situation occurs with respect to salinity. Since exposed tide pools reach high temperatures in the summer, evaporation is accelerated. As discussed in chapter 7, only the water evaporates, leaving the dissolved salts in solution, a circumstance that may increase the salinity drastically. In the winter, ice may form in the tide pool. This too increases the salinity, since dissolved solids are excluded from the ice crystal. When the tide rises, seawater of a lower salinity is brought into the tide pool. An exposed tide pool may also experience a large decrease in salinity during periods of rainfall, only to have the salinity increase rapidly as the tide pool is inundated with a rising tide. Thus, the biota of tide pools must also be able to withstand large and rapid salinity fluctuations.

Oxygen levels are also very variable in tide pools. Dissolved oxygen levels are temperature dependent, (chapter 7) and as the water temperature increases, the concomitant increase in molecular motion will disrupt the intermolecular forces that keep the oxygen in solution. For this reason as the temperature of a tide pool increases, the dissolved oxygen levels decrease. This may result in severe oxygen depletions during the summer. Tide pools containing large populations of plants are remarkably vulnerable to oxygen depletions during warm summer nights. Although the plants do photosynthesize and produce oxygen, photosynthesis occurs only during the day. At night the plants are incapable of carrying out photosynthesis, but they do respire along with the animals in the tide pool. This added respiration, in the absence of photosynthesis, often totally depletes a tide pool of dissolved oxygen. When the tide rises, the tide pool is once again inundated with oxygen-rich seawater.

Although the tide pools of the mid- and high-tide horizons retain water and protect the biota from desiccation, they do subject their inhabitants to other, very severe conditions. As a result, only those organisms that are capable of withstanding these conditions are found in tide pools: green algae, anemones, sea urchins, barnacles, whelks, etc.

Sandy Beaches

Sandy shorelines offer a distinct contrast to rocky beaches. In these areas the substrate is not stationary but is easily moved about by the currents and waves. This creates a continually shifting environment that offers few points of attachment for the plant and animal populations that are characteristic of rocky shorelines. In response to the constantly shifting substrate, burrowing organisms are the inhabitants of sandy shores; the surface sediments often appear to be, and indeed often are, devoid of life.

The major abiotic factor controlling plant and animal distributions on sandy shorelines is the sediment size. Sediment size directly influences the amount of interstitial water that is retained by the sediments. Fine sands, with their many and tortuous passageways, are able to retain large amounts of interstitial water. As a result the subsurface sediments seldom dry out, even when located in the high-tide horizon. Coarse sediments, on the other hand, are well drained and the subsurface sediments tend to dry shortly after the tide recedes. Consequently the burrowing organisms that inhabit substrates composed of fine sands are seldom vulnerable to desiccation. Furthermore overlying sediment provides a great deal of insulation; therefore, the interstitial water surrounding the organisms seldom reaches either lethal high or lethal low temperatures. Although the substrate is constantly shifting on sandy shores, the organisms inhabiting these areas are exposed to less stringent conditions than those of the rocky shorelines.

Since the substrate is continually shifting, the burrowing organisms have developed one of two major adaptations that enable them to survive under these conditions. Organisms such as the Pismo clam and "olive" snail of the Pacific coast burrow so deeply into the substrate that they are seldom affected by substrates that shift in response to normal wave and current action. These organisms have also evolved very heavy shells that aid in anchoring them in the substrate. Other organisms live closer to the

The olive snail burrows deeply into the sediment.

surface and are almost continually exposed when waves and currents strip away the overlying substrate. In response these organisms are able to rapidly rebury themselves. The razor clams of the east coast and the donax clam of the Gulf coast are examples of the rapid burrowers of sandy shores.

Since the macroscopic algae of rocky shorelines are absent on sandy substrates, significant grazer populations are also absent. The organisms that inhabit sandy shores are either detritus or suspension feeders (chapter 8). The few surface-dwelling animals pursue a scavenger life-style, feeding on detritus—the fine particles of plant and animal tissue cast upon the shoreline by waves. Razor clams, surf clams, and donax are all suspension feeders and filter plankton from the water during high tide.

Sand-beach communities, unlike rocky shorelines, do not have predator populations. This is due to the subterranean existence of much of the biota, which discourages predation and the presence of predators.

Mud Flats

Although mud flats are merely a variety of unconsolidated shoreline, composed of silt and/or clay-sized particles, there are sufficient biotic and abiotic differences to consider these areas separately. Since clay and silt are deposited only in areas of minimal water movement, mud flats develop only in areas that are protected from significant wave and current action. Since wave action is minimal, the slopes of these areas are much gentler than the slopes of sandy shores and are appropriately termed mud flats.

Due to the flat topography and very fine sediment size, mud flats are poorly drained and interstitial water is retained for long periods of time. In addition these areas have high populations of interstitial bacteria. The presence of the bacteria, in conjunction with the long residence time of the interstitial water, often results in the absence of oxygen in the interstitial water. Thus, oxygenless or anaerobic conditions are characteristic of mud flats.

Large populations of surface-dwelling mud snails, as well as burrowing animals, are present in mud-flat communities. The mud snails do not have to cope with anaerobic conditions, since they live on the surface sediments. These animals feed on the abundant detritus that is contained on and between the clay and silt particles and possibly feed on surface-dwelling bacteria as well.

The burrowing organisms must, however, contend with the anaerobic conditions of the subsurface sediments. Most are able to exist under these conditions by constructing burrows that have permanent openings on the surface of the mud flat, permitting air to move into the burrow. The vast majority of holes observed on mud flats, mark such burrows. The burrowing organisms are either deposit or suspension feeders. Most of the worms and some species of clam found on mud flats, are deposit feeders and obtain their food by ingesting the substrate, using the food contained thereon and excreting the unusable sediment. The suspension feeders, other species of clam and crabs, filter and feed on the plankton in the water. As on sandy shorelines, the burrowing life-style of the majority of the inhabitants of mud flats precludes the development of predator populations.

The Vegetation of Unconsolidated Shorelines

With the exception of the attached seaweeds of rocky shorelines, the major vegetation of the intertidal zone is marsh grass in the temperate zones and mangroves in the tropics. Both of these vegetation types develop on protected, unconsolidated shorelines, particularly on mud flats.

Salt marshes consist of two dominant plants: salt-marsh cordgrass, or *Spartina alterniflora,* and salt-marsh hay, or *Spartina patens.* Once an area accumulates sufficient sediment to be exposed to the atmosphere for at least a portion of the tidal day and the water depth at high tide is less than 2 meters (6

feet), cordgrass will rapidly colonize the area. Although this species is said to prefer freshwater environments, it is outcompeted in these areas by other species and is restricted to the intertidal zone, where the other plants cannot withstand the tidal flooding.

Cordgrass is able to withstand considerable immersion in estuarine water; it therefore colonizes the low-tide horizon, where interspecific competition is completely absent. Once the cordgrass moves into this area, it forms an effective barrier and slows the water on an incoming tide. This reduction in water velocity causes sediment to drop out of suspension and be deposited at the base of the cordgrass. As a result the marsh becomes higher, the period of tidal flooding decreases, and the marsh receives less water on rising tides. This forms an even more effective barrier to water inundation, and the sediment begins to be deposited at an ever-increasing rate, building outward from the seaward edge of the cordgrass stands.

Eventually, these seaward portions become sufficiently elevated to be inundated only at high tide. When this occurs the cordgrass moves into and colonizes these newly formed intertidal regions. In this manner the salt marsh moves seaward and encroaches on the open water by a combination of the physical deposition of sediment and the growth of the cordgrass. Eventually, the landward portions of the marsh originally colonized by the cordgrass will rise to such a height that they will no longer be flooded twice a day.

Although these high marshes provide a suitable habitat for the cordgrass, they are rapidly invaded by the smaller variety of *Spartina*—the salt hay. This species can only withstand the limited amount of tidal flooding that occurs bi-monthly, when the higher portions of the salt marsh are inundated by the spring tides (chapter 4). The salt hay, being a superior competitor, tends to eliminate the cordgrass from the high marsh. Thus, the cordgrass is effectively restricted to the portions of the marsh that are flooded daily and are therefore an unsuitable habitat for the salt hay.

It has been estimated that only 20 percent of either of the *Spartinas* are consumed directly by the grazers. Rather, in the colder temperate regions, these plants comprise a large standing crop that persists well into the autumn. In the winter these plants

The red mangrove is found in the low-tide horizon.

are mowed down when ice is pushed up onto the marsh. This ice-mown vegetation is then carried out into the estuary, particularly on the falling spring tides that inundate the highest portions of the marsh. The various marsh grasses either sink in the deeper portions of the embayment or are cast ashore. In either case they are moved about by wave and current action and ultimately are reduced to detritus and enter the scavenger food chain.

Salt marshes are quite common along the east and Gulf coasts of the United States. These areas have broad, gently sloping coastal plains, numerous barrier beaches and associated estuaries, a continental shelf, and a limited number of submarine canyons. All these factors, either singly or in conjunction with one another, are conducive to the formation of salt marshes.

The west coast, on the other hand, has steep coastal mountains that end at the shoreline. Consequently the few rivers have narrow mouths that severely restrict the extent of an estuarine environment. Moreover the narrow continental shelf is intersected with numerous submarine canyons that serve to drain off coastal sediment and carry it far offshore (chapters 2, 3). The result is that there are few barrier beaches along the coastline of the west coast and so, few salt marshes.

Mangrove forests, or mangal, are the tropical and subtropical counterparts of the salt marshes of the temperate zone. The term mangrove refers to the individual plant, while mangal denotes the entire community. Mangroves are actually terrestrial trees that have reinvaded the intertidal zone. They invade a protected shoreline once the normal sedimentation has raised it to a sufficient height to expose it at low tide. The process is similar to that described for salt marshes; the mangroves serve to trap additional sediment and thereby accelerate the depositional process and the development of the mangal.

Three species of mangrove are often found in a mangal, their positions determined by the degree of tidal inundation. The red mangrove is found in the

Black mangrove

All three species of mangrove are found in a mangal.

White mangroves are found at and above the high-tide horizon.

low-tide horizon and therefore can withstand a greater amount of tidal flooding. The black mangrove occupies the midtide horizon and can withstand less inundation. And the white mangrove is found at and above the high-tide horizon and can tolerate only a minimal amount of flooding.

Mangals trap large amounts of sediment and detritus. Hence, the bottom-dwelling organisms commonly found in these areas, such as the fiddler crab and the ghost crab, are detritus feeders. Both of these species live in burrows, which serve to carry oxygen to the deeper sediments, reducing the anaerobic conditions that tend to develop.

Barrier Beaches

Although barrier beaches are not, by definition, a part of the shoreline (chapter 12), they are important components of the coastal zone. The vegetation of

The mangroves trap large quantities of sediment.

barrier beaches is important, since it stabilizes and builds the beach. Barrier beaches, and particularly barrier islands, with significant stands of vegetation are less likely to breach and/or overwash.

The plants of the barrier beach must cope with two major factors. Barrier beaches are often windswept, and the winds generally carry sand that tends to scour the plants and either bury or blow away seeds before they are able to germinate. In addition the sand substrate of barrier beaches is well drained and holds little water; thus, there is very little fresh water available to the plants.

Once a barrier beach permanently emerges from the sea, rainfall washes the salts from the sediments, making them suitable for invasion by dune grass in the temperate zone and sea oats in the subtropic and tropics. Both of these plants are remarkably well adapted for life on an arid, windswept barrier beach.

They possess waxy, flexible leaves that bend and offer little resistance to the winds. The leaves do, however, slow the normal winds sufficiently to cause windblown sand to be deposited at the base of the plant, thus building the height of the beach.

These plants are able to reproduce both sexually, by producing seeds, and asexually, by sprouting from rootlike runners called rhizomes. Since seeds are either rapidly buried or blown away, the dominant form of reproduction is the latter. The rhizomes also serve another function: since they travel for considerable distances beneath the sand, they are able to obtain scarce water from a large portion of the beach. The runners must, however, be covered by several inches of sand before they are able to develop new shoots. This is a distinct disadvantage on many beaches, since human traffic removes the sediment and prevents the plants from reproducing.

Devegetated dune

As the dune grass or sea oats trap windblown sediment and raise the height of the beach, the landward portions of the barrier beach are afforded protection from the wind. This protection is generally sufficient to allow shrubs to begin to invade these areas. The shrubs tend to overgrow the dune grass or sea oats and outcompete them for light. Also shrubs such as beach plum reproduce by forming seeds and fruit, which tends to attract birds. A combination of bird droppings and decomposing fallen leaves at the base of the shrubs converts the sand substrate to soily sand. This type of substrate is able to hold a greater amount of rainwater.

The substrate continues to develop more soil, enabling it to hold an increasingly greater amount of interstitial water. Eventually sufficient soil will develop for trees such as the red cedar and pine to begin to move in, ultimately outcompeting the shrub stage.

In this way the more protected portions of the barrier island become a forest. The initial trees that invade such areas are often called pioneer species. The pioneer species also modify and build the soil, thereby making the substrate suitable for supporting other species of trees, generally various species of oak. These trees move into the area, outcompete the pioneer trees, and convert the area to an oak forest. Over the long term a combination of abiotic and biotic interactions serves to convert a barren, sandy barrier beach into a woodland.

Summary

The coastal zone provides an unrivaled opportunity to observe ecological relationships. The interaction of abiotic and biotic factors is graphically illustrated in the zonation of rocky shores, where wave shock and tidal range are reflected in the various species that inhabit these areas. The influence of substrate type is obvious when comparing the biota of rocky, sandy, and muddy shorelines.

The vegetational sequences are excellent indicators of abiotic factors on salt marshes and mangals. The presence of cordgrass and the red mangrove illustrates the extent of the inundation by the daily high tides, while the black mangrove and salt hay show the extent of inundation by the spring tides. Barrier beaches, on the other hand, illustrate the beach- and soil-building capabilities of the various vegetational species.

Review

1. What is a species?
2. Explain the concepts of habitat and niche.
3. Explain Gause's principle.
4. Distinguish between inter- and intraspecific competition.

5. What is the significance of the differential reproductive rates of various organisms within a population?
6. What is a biogeographic change point?
7. Define and give examples of biotic and abiotic barriers.
8. How does temperature affect the distribution of organisms in the intertidal zone?
9. How might the Coriolis effect affect biotic distributions in estuaries?
10. Compare and contrast the biota of mud flats, rocky shorelines, and sandy shorelines.

References

Futuyma, D. J. (1979). *Evolutionary Biology.* Mass.: Sinauer Associates.

Levinton, J. S. (1982). *Marine Ecology.* New Jersey: Prentice Hall.

MacArthur, R. H. (1972). *Geographical Ecology.* New York: Harper Row.

Stephenson, T. A., & Stephenson, A. (1972). *Life Between Tidemarks on Rocky Shores.* California: W. H. Freeman, Co.

For Further Reading

Gosner, K. L. (1971). *Guide to Identification of Marine and Estuarine Invertebrates.* New York: Wiley Interscience.

Johnson, M. E., & Snook, H. J. (1955). *Seashore Animals of the Pacific Coast.* New York: Dover.

Ricketts, E. F., & Calvin, J. (1968). *Between Pacific Tides.* California: Stanford Press.

14

The Ecology of the Open Ocean

Key Terms

abyssal zone
bathyl zone
benthic realm
competitive interference
food pyramid
hadal zone
inner neritic
neritic zone

oceanic zone
outer neritic
pelagic realm
pyramid of biomass
pyramid of energy
pyramid of numbers
sublittoral zone
trophic level

The organisms that inhabit the waters and sea bottom offshore the estuaries are subjected to a remarkably consistent abiotic environment. Tidal range is unimportant, and salinity and temperature exhibit a narrow range in any given area. Since the continental shelf and the abyssal plains are flat and, for the most part, featureless and since the abiotic environment is relatively uniform, there is little opportunity for a diversity of niches to develop. As a result the variety of organisms so evident in estuarine and intertidal habitats is, for the most part, lacking in offshore communities.

The Subtidal Zone

The subtidal zone is actually composed of two distinct components: those found in the water column itself and those that occupy the sediments. The water column is termed the **pelagic realm**, while the bottom is known as the **benthic realm**, or the benthos (fig. 14.1).

The pelagic realm may be subdivided horizontally into the **neritic** and the **oceanic zones**. The neritic zone encompasses all the waters that lie above the continental shelf. This zone is frequently subdivided into the **inner** and the **outer neritic** areas. The inner neritic portion includes all the waters immediately seaward of the low-tide horizon to a water depth of 50 meters (160 feet). The 50-meter mark is considered to be a convenient water depth at which to divide the neritic zone, since it is generally the maximum depth to which sunlight can penetrate into the water column. Thus, the bottom sediments beneath the inner neritic area fall within the euphotic zone. At depths greater than 50 meters, phytoplankton respiration exceeds photosynthesis and plants cannot establish permanent populations. The outer neritic area includes all the waters above the continental shelf that have a depth of 50 meters or more. This zone extends to the continental slope. The oceanic zone encompasses all the waters offshore the continental shelf.

Box 14.1 Plants at 290 Meters

Plants have recently been found at water depths of over 290 meters (950 feet) in the Caribbean Sea. These plants were found on the side of a ridge off the northern border of San Salvador Island in the Bahamas.

A previously unknown type of algae associated with coral, these plants have evolved thickened side walls and very thin upper walls. Apparently the structures that are involved in photosynthesis are lined up vertically to enable them to capture the very low light at these great depths. The alignment of these structures enables the algae to survive and photosynthesize in depths that contain only one one-hundredth of the light required by marine plants that inhabit lesser depths.

The pelagic realm may also be divided vertically on the basis of the penetration of sunlight into the water column (fig. 14.2). This is akin to the zonation in lakes (chapter 7). The upper waters of effective sunlight penetration are termed the euphotic zone, while the aphotic zone includes all the waters beneath the zone of effective sunlight penetration.

The benthic realm is subdivided on a horizontal basis into the **sublittoral**, the **bathyl**, the **abyssal** and the **hadal** zones (fig. 14.3). These divisions of the benthos are based upon the water depth that lies above each particular zone. The sublittoral zone consists of all of the sediments and associated animals and plants that occupy the continental shelf. It therefore encompasses the area immediately seaward of the low-tide horizon to the continental slope.

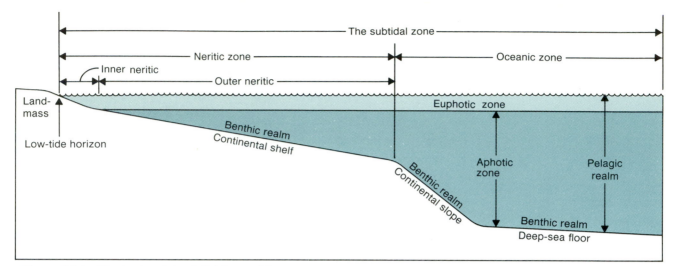

Figure 14.1 The horizontal subdivisions of the subtidal zone

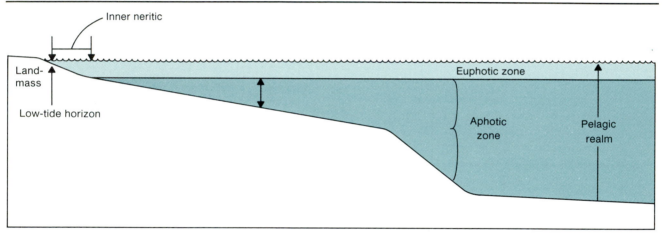

Figure 14.2 The vertical subdivisions of the pelagic realm

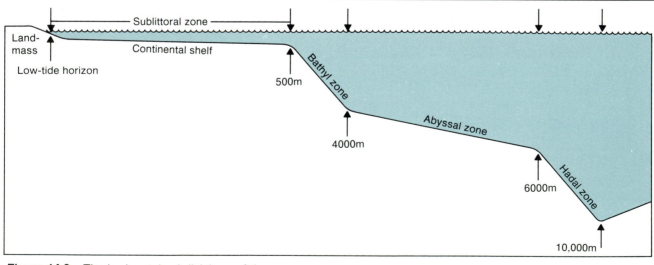

Figure 14.3 The horizontal subdivisions of the benthic realm

Eel grass is found in shallow bays of the east coast.

Turtle grass replaces eel grass in the shallow tropical seas.

The depth of water overlying this zone ranges from just millimeters to approximately 500 meters (1,600 feet). The waters of the inner neritic zone of the pelagic realm overlie only the landward portions of the sublittoral zone. Although only a small portion of the sublittoral benthos is within the realm of effective light penetration and is able to support benthic plant populations, the plants are so abundant in these shallower areas that the communities are often named for their dominant plant populations, since those are the most prominent features of these regions. Hence, the terms eel-grass community, kelp beds, and turtle-grass community are frequently used to designate specific communities. As was the case with the designations used for the intertidal zone (chapter 13), the terms indicate the types of organisms that are found in these communities.

Turtles feed extensively on turtle grass.

The vast majority of the sublittoral zone is, how-ever, far below the euphotic zone, and benthic plants are absent. As a result the grazers are also absent from these communities. The animals that occupy these deeper regions of the sublittoral benthos must pursue scavenging life-styles.

The bathyl zone lies beyond the sublittoral and consists of those bottom areas of the continental slope that lie beneath 500 to 4,000 meters (1,600 to 13,000 feet) of water. This zone is obviously far beneath the euphotic zone, and the biota consists solely of animals and bacteria.

The abyssal zone extends beyond the conti-nental slope to water depths of 6,000 meters (19,500 feet). This zone encompasses the abyssal plains of the deep-sea floor (chapters 2, 3). The benthos of the trenches that intersect the deep-sea floor is termed the hadal zone. The depths of the hadal zone range between 6,000 and 10,000 meters (19,500 to 33,000 feet).

The Ecology of the Continental Shelf

With the exception of estuarine organisms and those that inhabit the portions of the subtidal zone im-mediately adjacent to landmasses, the animals and plants that inhabit the pelagic and benthic realms of the continental shelf are exposed to relatively constant abiotic factors. For example, these organisms are never exposed to the atmosphere. As a consequence desiccation and the possibility of encountering lethally high and/or lethally low temperatures is not a problem in these offshore areas. Since these organisms inhabit the waters and benthos that are seaward of barrier islands, sand spits, bay-mouth bars, etc., neither are they exposed to the widely fluctuating salinities nor to the wide temperature ranges characteristic of estuarine waters and those areas in the vicinity of river mouths.

The most important abiotic factors that influence the structure of subtidal communities are sunlight and the turbulence caused by waves and currents. In the polar seas and, at times, in the colder temperate regions, sea ice may affect the structure of subtidal communities.

Since sunlight is able to penetrate to the bottom of the inner neritic zone, this region is entirely within the euphotic zone and is the only portion of the pelagic realm that contains phytoplankton populations throughout the water column. Moreover, since the underlying benthos is also within the zone of effective light penetration, it is capable of supporting significant populations of macroscopic, bottom-dwelling plants.

As noted these bottom-dwelling plants are so abundant that they are often used to denote the community. For example, the kelps are found on rocky subtidal bottoms in the temperate zone of both the Atlantic and Pacific, while on the softer bottoms the sea grasses predominate in both the Atlantic and Pacific oceans. In the temperate zone of both these oceans, eel grass is the common sea grass; in the tropical regions the eel grass is replaced by turtle-grass communities.

These sea grasses serve as a food source for a variety of grazers. However, like *Spartina*, large quantities of these plants remain unconsumed, die, and are reduced to fine particles by mechanical grinding. This detritus then serves as an energy source for subtidal scavengers. Waves also cast a portion of this material onto the beach in the intertidal zone, and the detritus is then used by the scavengers of that zone.

Kelp and sea-grass beds are remarkably productive. For example, the kelp beds of California are estimated to produce 80 grams of usable material for marine food chains per square meter of kelp bed per year, while the eel-grass beds of the Atlantic Ocean produce between 500 and 1,000 grams per square meter annually. Thus, these benthic, macroscopic plants add significant quantities of food to the coastal ocean and, in conjunction with the phytoplankton of the estuaries and the inner neritic zone, make the coastal seas highly productive.

Turbulence plays a major role in maintaining the high phytoplankton populations of the neritic zone. As in estuarine areas, the turbulence prevents the waters of both the inner and outer neritic areas from forming temperature and/or salinity density barriers (chapters 7, 8, 9). Hence, dissolved nitrate and phosphate seldom become spatially unavailable to the phytoplankton that inhabit the surface waters. This is particularly important in the outer neritic zone, which is sufficiently deep and has an aphotic zone. The abundance of phytoplankton sustains high populations of zooplankton and other grazing organisms, which, in turn, serve as a food supply for other consumers. Thus, well-developed food chains evolve, based upon the phytoplankton.

The turbulence of these shallow areas stirs up and suspends the finer sediments, primarily the clay- and silt-sized particles. These materials are then transported to and deposited in the calmer waters of an estuary, or they may be carried to and ultimately deposited in the deeper waters farther offshore (chapters 2, 5). As a result the bottom sediments of the inner neritic zone are generally composed of sands, pebbles, or rocks. Regardless of the sediment type, however, the benthos of the continental shelf is, for the most part, flat and featureless (chapter 2). Thus, there is little variety in these habitats and little diversity of niche. Most of the benthic animals are either deposit or suspension feeders. The deposit feeders utilize the detritus on or in the sediments, while the suspension feeders obtain their energy from the plankton and/or detritus that is suspended slightly above the bottom.

The major inhabitants of the subtidal zone are various species of burrowing worms and molluscs and surface-dwelling brittle stars, sand dollars, and sea urchins. The burrowing organisms are most abundant in the muddy sediments, while the suspension feeders are primarily found on sandy bottoms.

With the exception of the ice-induced mortality of the biota of rocky shorelines (chapter 13), ice does not generally affect the distribution of the animals and plants in the subtidal portion of the temperate zone. In the polar seas, however, ice does play a significant role in influencing the biota of the benthos that underlies the inner neritic zone.

In these areas several meters of ice form during the colder months. At times this ice extends down to and meets the bottom sediments. When this occurs, benthic organisms, particularly the surface dwellers

Red abalone

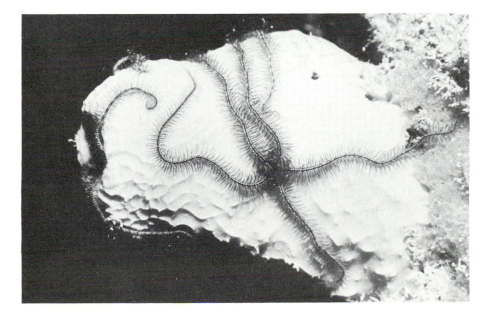

Brittle star

Box 14.4　Atmospheric Debris and Extinction

Ammonite

The reproductive strategy developed by a given species may play a significant role in its success or extinction. Two groups of marine molluscs, the ammonites and the nautili, occupied essentially the same habitats in the world's ocean 70 million years ago. The ammonites were much more common during this time. For example, in some areas sediments containing fossils of both groups are estimated to contain hundreds of millions of ammonites but only thousands of nautili. Yet today all of the ammonites are extinct, while the nautili are still present.

This mass extinction occurred 65 million years ago and was accompanied by the extinction of 90 percent of the species of calcareous plankton. It is believed that this was due to an asteroid collision that produced atmospheric debris. The debris would have reduced or totally eliminated sunlight, which in turn, would have inhibited photosynthesis. As a result large numbers of phytoplankton would have been killed. The secondary consumers, such as merozooplankton, would have been deprived of their food supply and also have been killed.

The ammonites are known to have produced very small eggs that hatched rapidly. The larval ammonites then would have spent a long period as zooplankton before assuming their adult form and

Nautilus

life-style. The nautili, on the other hand, produce large eggs and when the young hatch, they are active swimmers rather than passive zooplankton. An adult mode follows shortly. In other words the small eggs produced by the ammonites predetermined the young to spend a lengthy planktonic existence. The large eggs of the nautili resulted in a very short planktonic stage, followed by an active swimming stage, and then the adult life-style.

The ammonites, with a prolonged planktonic stage, would have been affected to a much greater extent by the large-scale death of the phytoplankton. Since they are now extinct, this may indeed have been the case.

The release of atmospheric debris by asteroid collision would be akin to what would occur on a much larger scale in the case of nuclear war. Vast quantities of radioactive debris, smoke, and so forth would be released to the atmosphere and eliminate sunlight penetration to the earth. The resulting "nuclear winter" would have a disastrous impact on those organisms that survived the nuclear attack. Since air currents would spread this debris throughout the atmosphere, the entire earth would experience this nuclear winter.

may be killed by ice grinding or freezing. Ice-induced mortality also occurs in the intertidal zone, generally to a greater extent. As a result permanent benthic communities do not develop from the high-tide horizon to the seaward limit of this ice.

Anchor ice, a term given to ice that actually forms on the bottom, presents another barrier to the permanent habitation of the benthos in the polar seas and, occasionally, in the colder portions of the temperate zone. Anchor ice forms directly on the bottom when the water temperatures are very cold and water crystallizes about a convenient object, termed a nucleus. When this occurs additional ice crystals form and develop into platelets. As the platelets increase in size, they eventually surround, engulf, and freeze sessile organisms.

Predation is an important biotic factor that influences the composition of benthic communities. Various fish, as well as crabs, starfish, and predatory snails prey upon benthic organisms.

In many areas the life-styles of the benthic inhabitants tend to influence the structure of their communities. For example, the burrowing organisms appear to discourage the presence of suspension feeders in some areas, while the suspension feeders discourage the presence of burrowing animals in other areas. Apparently the deposit-feeding burrowers stir the sediments and make them unstable. The slightest water movement will resuspend these sediments, which would then clog the feeding apparatus of the suspension feeders. In addition as the burrowers move the sediments about, they tend to bury and thereby kill the newly settled larvae of the suspension feeders. Larval-deposit feeders, however, burrow into the deeper, more stable sediments and are unaffected by the shifting surface sediments. The suspension feeders, in turn, filter and feed upon the larvae of the burrowers when they are in their suspended stage and are located slightly above the bottom. The exclusion of various populations by the interference with their life-styles or reproductive strategies is termed **competitive interference.** Competitive interference may have profound effects on the distribution of benthic populations.

The Ecology of the Oceanic Zone and Associated Benthos

Whereas a considerable portion of the waters in the neritic zone fall within the euphotic zone, the vast majority of the oceanic zone is in perpetual darkness. In fact it is estimated that 85 percent of the total area of the sea is in the aphotic zone, which constitutes 90 percent of the total volume of the world's seas. Thus, only 15 percent of the total area of the sea, from the intertidal zone to the euphotic zone of the deep ocean, is capable of supporting plant life.

The upper, sunlit waters of the sea contain phytoplankton, and the typical grazer-based food chains develop. In many of these areas, evaporation tends to increase the salinity at the surface, causing the surface waters to become denser and sink. This results in a net downward movement of dissolved phosphate and nitrate. Since these offshore seas are deep, turbulence does not tend to mix the surface and subsurface waters, and the nutrients may be trapped below a density barrier beneath the euphotic zone. The spatial unavailability of nutrients is quite common in the oceanic zone, and once this occurs, the nutrients may be transported for considerable distances by the deep-ocean currents before being returned to the euphotic zone (chapters 9, 11). With the exception of the spatial unavailability of nutrients and the absence of macroscopic benthic plants, the ecology of the euphotic zone of the pelagic realm is similar to that discussed for the neritic zone. The unique portions of the pelagic realm are the waters of the aphotic zone.

Light, pressure, salinity, temperature, and dissolved oxygen are relatively uniform in any given area of the oceanic zone, though they may vary widely from area to area. For example, light levels are continually high in the euphotic zone of the tropical ocean, but light is totally absent in the aphotic zone of this same sea.

Both temperature and salinity are remarkably constant. Except in the tropical seas, salinity is generally lower at the surface and increases with depth. This is to be expected, since salinity directly affects the density of water, and the denser, more saline water would tend to sink. Although there are slight salinity variations in any given area, they are not ecologically significant. Temperature decreases gradually with depth in the euphotic zone and more rapidly in the thermocline (chapter 7), which is considered to be the transition zone between the surface and the deeper waters. At water depths of 3,000 meters (9,800 feet) and greater, the waters are homothermous. In these waters there are neither seasonal nor annual temperature changes. In fact, the temperatures of the deep ocean are remarkable by their constancy. Of all of the earth's environments, only the deep ocean, caves, and hot springs have such constant temperatures.

Although pressure is of minimal significance in estuarine areas and in the inner neritic zone and associated benthos, it is the most variable abiotic factor in the deep ocean. At sea level an organism is exposed to a pressure of 1 atmosphere or 14.7 pounds per square inch. The pressure increases by 1 atmosphere for every 10 meters (32 feet) in depth. Thus, a benthic organism of the abyssal zone is under waters ranging from 4,000 to 6,000 meters (13,000 to 19,500 feet), and is exposed to pressures from 400 to 600 atmospheres, or from 688 to 882 pounds per square inch. Since the majority of the deep ocean ranges from 2,000 to 6,000 meters (6,500 to 19,500 feet) in depth, most of the deep sea and the associated organisms are exposed to pressures between 200 and 600 atmospheres. No other environment has such a high or extreme range of pressures. Few macroscopic living organisms have been retrieved from the deep ocean; therefore, little is known about the adaptations that enable them to live under extreme pressures. Deep-sea bacteria have been obtained but fail to grow or reproduce at lower pressures. Obviously these marine organisms must have remarkable adaptations that enable them to survive at these depths. Organisms living at a particular depth are adapted to the pressures at that depth. Thus, the pressure, per se, is not a problem. Problems do arise when the organisms change depth and encounter different pressures.

Since light is completely absent in 85 percent of the world's seas, living plants cannot exist nor can photosynthesis occur. As a result the organisms of the deep ocean are totally dependent on the euphotic zone as a primary energy source. This food must sink beneath the upper, densely populated waters, and much, if not all, is consumed before it can reach the deeper levels. In all likelihood the only materials that reach these waters are particles such as fecal pellets, bones, dissolved material, and the protective outer coverings of organisms that are not directly usable as a food supply. These materials must be broken down by the benthic bacterial populations and converted into bacterial tissue before they can be used by the organisms of the deep sea. Larger predators then, presumably, feed on the bacterial grazers. Thus, grazer food chains, based upon bacteria rather than phytoplankton, develop in the deep ocean. This is similar, if not identical to, the food relationships that develop in the Galapagos Rift zone (chapter 2).

Other sources of food in the deep ocean are animals such as squid that spend their juvenile periods in the euphotic zone and then move to the aphotic zone as adults, where they may serve as a food supply for deep-sea predators. The bodies of large marine mammals such as dolphins and whales sink so rapidly that they are not completely consumed by the inhabitants of the shallower waters. The unconsumed portions also provide a food supply for the deep-sea organisms.

In general, however, food is scarce in the deep ocean, a fact that is reflected in the low population densities of these areas. Another indication of the paucity of food is the small size of the benthic organisms, particularly those of the abyssal and hadal zones. These organisms—primarily worms, crustaceans, and molluscs—are much smaller than their counterparts that inhabit the shallower seas.

Another indication of the low animal populations in the deep sea is afforded by the oxygen levels in these areas. Oxygen is relatively abundant in this portion of the aphotic zone of the deep ocean. As noted in chapter 11, large amounts of dissolved oxygen are carried into and distributed throughout the deep ocean by the Arctic and Antarctic water masses as these waters descend through the less dense waters of the temperate zone as pressure-gradient currents. Curiously, however, oxygen levels decline significantly at water depths between 500 and 1,000

Box 14.5 Bottom Sampling Devices

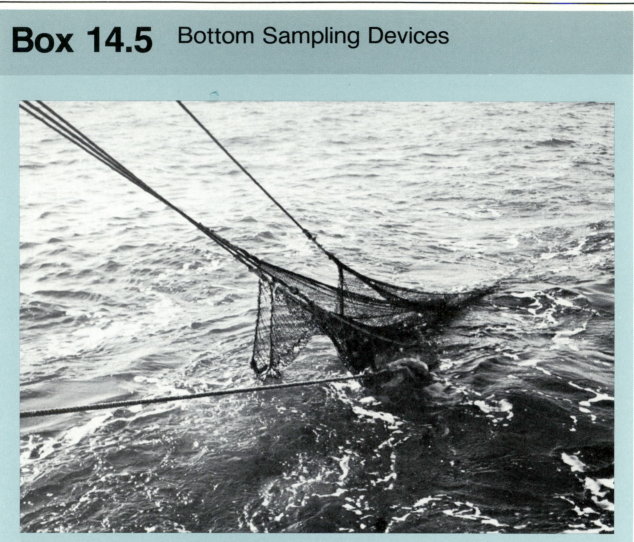

Subtidal organisms are commonly collected by using a bottom trawl.

Benthic organisms are generally collected by using dredges, grab samplers, or bottom trawls. A dredge is a boxlike apparatus that is lowered to the bottom and dragged behind the ship. As the dredge moves over the bottom, it scrapes organisms from the surface and sweeps them into the open-ended dredge box. A wire- or cloth-mesh liner is placed inside the dredge box to permit a free flow of water, but to retain the organisms. The mesh size of the liner determines the size of the organisms that are collected.

A grab sampler consists of a pair of jaws and an upper reservoir to hold the sample. The grab is released from the ship with the jaws open. Where it strikes the bottom, the jaws "trip," snap shut, and obtain the sample.

Bottom trawls are often used instead of dredges and grabs to obtain larger benthic organisms. A typical bottom trawl consists of large mesh nets that drag along the bottom and collect whatever is in their path.

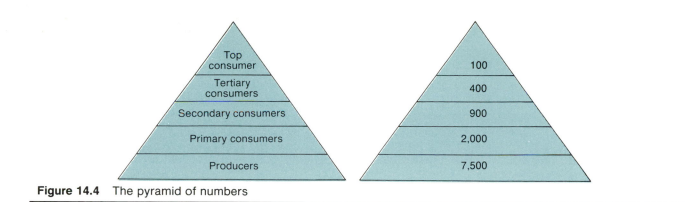

Figure 14.4   The pyramid of numbers

meters (1,600 to 3,200 feet), causing this area to be termed the zone of the oxygen minimum. This zone is at the edge of the continental shelf and is known to contain large amounts of food. In response the animal and bacterial populations are also high, and these organisms remove large amounts of oxygen as they respire. Beneath this zone oxygen levels remain higher, since the animal populations, and hence respiration, are low.

Trophic Levels and Food Pyramids

In addition to depicting plant–grazer–consumer relationships by food chains, they may also be illustrated by constructing **food pyramids** based on numbers, biomass, or energy. This is done by assigning all the inhabitants of a given community to a particular **trophic level,** (chapter 9) which is defined as the level at which an organism obtains its food and therefore its energy.

Since plants obtain their energy from the sun, they are assigned to the first trophic level. The grazers, such as the copepods (chapter 8), obtain their energy by feeding on the plants, and hence occupy the second trophic level. Those animals that prey upon the grazers are functioning at the third trophic level, and so forth.

Once these assignments have been made, the various pyramids can be depicted. The **pyramid of numbers** (fig. 14.4) is constructed by actually counting or estimating the number of plants and primary, secondary, tertiary, etc., consumers that are present per unit volume of the community. The

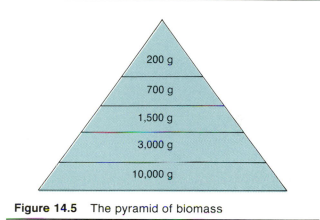

Figure 14.5   The pyramid of biomass

number of plants are assigned to the first trophic level and are placed at the base of the pyramid. The primary consumers—the grazers—are placed directly above the plants, since they feed on the plants. The secondary consumers are placed above the grazers, and so forth.

The **pyramid of biomass** is constructed in a similar manner. Instead of counting or estimating the number of organisms, a representative sample is taken, and the organisms that occupy each trophic level are dried and weighed. The dry weight of the plants per unit area is then placed at the base of the pyramid, the dry weight of the grazers per unit area above this, and so forth (fig. 14.5).

The **pyramid of energy** is derived by sampling each trophic level and determining the number of calories that are contained in the organisms at each level. This data is then used to determine the flow of energy from level to level (fig. 14.6).

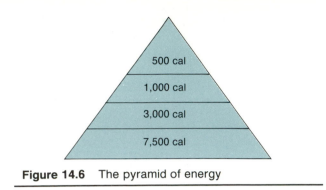

Figure 14.6 The pyramid of energy

In the sea the pyramids of numbers and biomass are frequently inverted as a result of the cyclic nature of phytoplankton blooms and the large size differences of the organisms at different trophic levels. The pyramid of energy, on the other hand, must always be upright. This is due to the loss of energy with each transfer and is in accordance with the laws of thermodynamics (chapter 6). Since energy is lost as it is passed from producer to grazer to predator, there must be a decrease in numbers, biomass, and energy at each higher level on the pyramid.

The Sea as a Food Supply

Since the vast majority of the sea is well below the euphotic zone, the plants that govern the sea's productivity are confined to an extremely small area in the deep oceans of the world. The productive portions of the oceans are very limited; in fact, they are so limited that it is a mistake to assume that the sea holds the solution to man's food needs.

In order for any animal population to increase, it must have sufficient food. The second law of thermodynamics, however, sets a definite upper limit on the animal populations in the sea. The higher the position of a population on a pyramid, the less the energy that is available to those animals.

Nations that have high human populations and limited agricultural land must feed close to the bottom of the pyramid. For example many Asian countries obtain a large portion of their energy requirement from plants. These human populations can only attain and maintain a high population density if they resort to a grazer type of life-style.

When western cultures speak of food from the sea, however, they envision lobsters, swordfish steaks, etc. All these foods are at or near to the top of the energy pyramid, and the second law of thermodynamics severely limits the number of organisms that occupy these trophic levels.

In order for the sea to be able to support truly large numbers of humans, man would be forced to feed lower on the sea's food pyramid. The vast majority of marine plants are not large and esthetically pleasing like lettuce, rice, and corn; rather, they consist of the phytoplankton. When these plants are filtered from seawater, they form a mushy, green mass. It is doubtful whether western cultures are yet ready to consider this as a suitable food supply.

Summary

The sea, offshore the inner neritic zone, is remarkable in its constancy. In any given area, the salinity and temperature are virtually invariable. Pressure, too, is constant, but in no other environment are the organisms subjected to such tremendous pressures.

The vast majority of the pelagic and benthic realms are in perpetual darkness, and effective photosynthesis does not occur. All the organisms of the open ocean, then, are dependent on that small portion of the sea that is in the euphotic zone. As a consequence the animal life is sparse, and very different food chains develop. In the deep ocean much of the initial energy is thought to be provided by the bacteria, which, like the phytoplankton of the euphotic zone, serve as a food source for the grazers.

The second law of thermodynamics, in conjunction with the small area that is actually inhabited by the producers, shows the fallacy of relying on the sea as a food supply. Should this become a reality due to overpopulation, it will be necessary to feed at a trophic level much closer to the energy source.

Review

1. Sketch the subtidal zone and subdivide it horizontally and vertically. Label all subdivisions.
2. What are the major abiotic factors that influence the structure of subtidal communities?
3. What is competitive interference? Give examples.
4. What are the major food sources in the deep ocean?
5. What is the significance of the zone of the oxygen minimum?
6. Explain how the three food pyramids are constructed.
7. Can any of these food pyramids be inverted? Why?
8. If a food pyramid is permanently inverted what does it indicate? Why?
9. Estimate how each of these food pyramids would be shaped for your community. Why would they take these shapes?
10. Why is it an error to rely on the sea as an answer to the food needs of the human population?

References

Levinton, J. S. (1982). *Marine Ecology*. New Jersey: Prentice Hall.

Moyle, P. B., & Cech, J. J. (1982). *Fishes: An Introduction to Icthyology*. New Jersey: Prentice Hall.

Reish, D. J. (1969). *Biology of the Oceans*. California: Dickenson Publishing Co.

For Further Reading

Issacs, J. D. (1969, September). The nature of oceanic life. *Scientific American*.

Sumich, J. L. (1984). *An Introduction to the Biology of Marine Life*. Iowa: Wm. C. Brown Publishers.

a

abiotic factor   The nonliving component or components of an ecosystem.

abyssal clay   Deep-ocean sediment containing less than 30 percent biogenic material.

abyssal hills   Low hills rising from the deep-ocean floor.

abyssal zone   That area of the deep ocean that extends beyond the continental slope to a water depth of 6,000 meters.

active continental margin   Sites of geological activity.

anchor ice   Ice that forms on the sea floor.

anadromous fish   Marine fish that return to freshwater areas to spawn.

aphotic zone   The zone of perpetual darkness.

apogee tide   The tide that is formed when the moon is at its farthest point from the earth.

asthenosphere   The more fluid, outer portion of the earth's mantle.

atoll   A ring-shaped coral reef that encloses a lagoon and is surrounded by the open sea.

autotroph   Organisms capable of producing food from carbon dioxide, nitrate, and phosphate.

azoic   Devoid of life.

b

back radiation   The radiation of heat from the earth back to the atmosphere.

backshore   The area of a beach that extends from the high-tide horizon landward to a cliff, dune, or zone of permanent vegetation.

bar-built estuary   An estuary protected by a bay-mouth bar.

barrier beach   A bay-mouth bar, sand spit, or barrier island that provides protection for the mainland.

barrier island   A long, narrow wave-built island separated from the mainland by a lagoon, bay, etc.

barrier reef   A coral reef parallel to but offshore an island or terrestrial landmass.

basalt   Black or greenish-black igneous rock.

bathyl zone   The bottom areas of the continental slope.

bay-mouth bar   Permanent, abovewater features, generally attached to a headland and extending across the mouth of a harbor, inlet, etc.

beach face   The section of the foreshore that slopes seaward and is exposed to wave action.

bed load   Large materials that are moved along a river bottom or bed.

benthic organism   Bottom-dwelling organisms.

benthic realm   The component of the subtidal zone that consists of the sea's bottom.

biogenic sediment   Sediment derived from animal and plant materials.

biogenic sorting   Sediment moved about and sorted by the life processes of organisms.

biogeochemical cycle   The process by which necessary materials cycle from the abiotic to the biotic and back to the abiotic environment.

biogeographic change point   Areas when there is an abrupt and obvious change in the biota. These changes are often due to abiotic factors such as temperature variations.

biological magnification   The process by which materials are taken in and accumulate in the tissues of organisms. As this material passes up the food chain, its concentration increases in the tissues of the higher consumers.

biological oxygen demand   Condition resulting from the removal of oxygen from a system due to bacterial processes.

biotic factor   The living component or components of an ecosystem.

breakwater   A structure of rock or concrete built offshore the surf zone and parallel to the shoreline.

c

caloric   The amount of energy required to raise the temperature of 1 gram of water 1°C.

calcareous ooze   Sediment containing over 30 percent of biogenic material in which the element calcium predominates.

carnivore   An organism that utilizes animals as a major or sole food source.

centrifugal bulge   The accumulation of water that is found on the side of the earth opposite the moon.

chemical unavailability   A condition in which necessary materials such as nitrogen and phosphorous are not in the proper form to be used by plants.

Glossary

chemical weathering   The tendency of certain materials contained in sediments to dissolve in water.

coastline   The area that extends along the shore and encompasses the area from the neap-tide mark to the landward limit inundated by storm waves.

cocolithophore   A phytoplankton covered by calcareous discs.

community   Groups of organisms of different populations inhabiting a given area.

compensation depth   The depth at which photosynthesis equals respiration.

competitive exclusion principle The principle that states that competition will always occur when two or more species occupy the same niche at the same time.

complex inlet   A situation that occurs when the updrift side of a barrier beach overlaps the downdrift side. The channel runs parallel to the barrier beach.

contact conduction   The transfer of heat directly from the sea's surface to the air.

continental crust   The crust composed of granite and its derivatives that forms the continental land masses.

continental drift   The movement of the continents over the earth's surface.

continental margin   The area consisting of the continental shelf and the continental slope.

continental rise   The gentle slope at the foot of the continental slope.

continental shelf   The portion of the sea floor that is adjacent to the continents and that extends offshore to the continental slope.

continental slope   The steep, downward slope at the seaward edge of the continental shelf. The continental slope marks the seaward extent of the continental landmasses.

contour current   A bottom current that flows parallel to the base of the continental slope.

convection   The movement of a fluid or molten substance away from the source of heat.

converging boundary   The boundary formed by plates that are moving toward each other.

copepod   A type of zooplankton.

core (of earth)   The central portion of the earth.

Coriolis effect   The tendency of water to appear to arc to its right in the Northern Hemisphere and to its left in the Southern Hemisphere.

cosmogenous sediment   Sediment derived from outer space.

covalent bond   A chemical bond formed by the sharing of electrons.

crust   The outer portion of the earth (*see* continental crust).

crustal rebound   The upward movement of the crust in response to the removal of weight.

d

decomposers   Organisms, primarily bacteria, that utilize plant and animal remains and waste products as an energy source.

deep-water wave   Any wave with a ratio of water depth to wave length greater than 1:2.

delta   A low-lying sediment deposit often found at the mouth of a river.

density   The relationship of an object's mass to its volume.

depositional shoreline   A shoreline that is gaining sediment.

detritus   Finely ground animal and plant remains.

diatom   A type of phytoplankton covered by valves composed of silica.

differential solubility   The tendency of some materials to dissolve to a greater or lesser extent than other materials.

diffraction   The lateral propagation of wave energy.

diurnal tide   One high and one low tide per day.

diverging boundary   The boundary formed by plates that are moving away from each other.

doldrums   A belt of light, variable winds located at 10° to 15° north and south of the equator.

dust-cloud hypothesis   The theory that holds that the planets were formed from a diffuse cloud or clouds of dust and gases.

dysphotic zone   The zone of perpetual darkness (*see* aphotic zone).

e

ecosystem   A group of interacting communities.

Ekman spiral   The movement of bottom water in the opposite direction from the surface waters. This, theoretically, occurs as a result of the Coriolis effect.

electron   A negatively charged particle found orbiting about the nucleus of an atom.

El Niño   A wind-driven reversal of the Equatorial Current in the Pacific Ocean. This reversal results in the shoreward movement of water toward the coast of South America.

emergent shoreline   A shoreline formed by a rise of submarine areas relative to the seas surface.

epilimnion   Zone of gradual temperature change.

erosional shoreline   A shoreline that is losing sediment.

essential element   Elements that are required by all animals and plants.

estuary   A semi-enclosed body of water with a free connection to the open sea. Such areas generally have large inputs of fresh water.

euphotic zone   Zone of sunlight penetration.

euryhaline organism   Organisms capable of tolerating a wide range of salinities.

evaporation   The physical process whereby a liquid is converted to a gas.

f

fault   Fractures in the crust where one side has been displaced relative to the other side.

fetch   Any area of the sea over which the wind blows with a constant speed and duration. The fetch is one of the factors that determines the height to which waves may build.

fissure   Tensional cracks in the earth's crust.

food chain   The transfer of energy (food) from producers to consumers to decomposers.

food web   A series of interrelated food chains.

foraminiferal ooze   Sediment in which foraminifera or their tests comprise over 30 percent.

foreshore That portion of a beach that is periodically covered at high tide and exposed at low tide.

fracture zone *See* fault.

freezing-point depression The lowering of the freezing point of a liquid due to the presence of dissolved materials that hinder hydrogen bond formation.

fringing reef A coral reef attached directly to an island or continental landmass.

g

Gause's principle *See* competitive exclusion principle.

geostrophic current The motion of water due to the Coriolis effect and gravity.

granite A light-colored igneous rock of the continental crust.

grazer An organism that feeds solely on plants.

groin A structure of stone or concrete extending from the beach into and seaward of the surf zone.

gyouts Sea mounts with flattened tops.

h

habitat The area actually inhabited by an organism.

hadal zone A subdivision of the benthic realm that consists of the trenches of the deep-sea floor. The depth of the hadal zone ranges between 6,000 and 10,000 meters.

heat of evaporation The number of calories required to evaporate water.

heat of fusion The number of calories that must be removed from water in order to freeze it.

hermaphrodite An organism that contains functional male and female reproductive organs. Rarely, however, is such an organism capable of self-fertilization.

heterotroph Organisms incapable of synthesizing carbohydrate from simple low-energy materials.

high-tide horizon That area of the intertidal zone that is inundated at high tide.

holoplankton Either phytoplankton or zooplankton that spend their entire lives as plankton.

homothermous Having a uniform temperature.

horse latitude A belt of light, variable winds located 30° to 35° north and south of the equator.

hurricane A tropical disturbance in which winds may reach velocities of over 160 km/hr (100 mph). The term *hurricane* is generally applied to storms in the North Atlantic, eastern North Pacific, the Gulf of Mexico, and the Caribbean Sea. In the western Pacific these storms are termed typhoons.

hydrogenous sediment Sediment formed by materials that precipitate from sea water.

hypolimnion Bottom water (in a lake) at a constant temperature of 4°C.

i

igneous rock Rock formed by the solidification of magma.

inner core The solid inner portion of the earth's core.

inner neritic zone A subdivision of the pelagic realm that encompasses all of the waters immediately seaward of the low-tide horizon to a water depth of 50 meters.

intermolecular attractive force The attractions that develop between molecules.

interspecific competition Competition between organisms of different species.

interstitial water The water found between sediment particles.

intertidal zone The area that is alternately covered and exposed during high and low tides.

intramolecular attractive forces The attractions that develop between atoms within the same molecule.

intraspecific competition Competition between organisms of the same species.

ion A charged particle.

ionic bond A chemical bond formed by the complete gain and loss of electrons.

isostacy Equilibrium.

isostatic adjustment The movement of the earth's crust in order to restore isostacy (equilibrium).

j

jetty A structure of stone or concrete constructed at the mouth of an inlet, harbor, etc.

juvenile water Water entering the sea from the earth's interior in rift areas.

l

lagoon A shallow body of marine water partially or completely separated from the open sea by a barrier island, barrier reef, etc.

latent heat The number of calories required to change the state of a substance at a given temperature and pressure.

latitude Lines drawn about the earth beginning at the equator and progressing northward and southward to the poles.

limiting factor Any factor that limits the success of a population.

lithogenous sediment Sediment that is derived from rocks.

lithosphere The continental and oceanic crust, as well as the upper mantle.

longitude Lines drawn from pole to pole intersecting the lines of latitude.

long-range order Large aggregations of molecules held together by strong intermolecular attractive forces.

longshore current A wind-induced current moving in the surf zone.

loop The point on a standing wave where the water moves vertically.

low-tide horizon That portion of the intertidal zone that is exposed at low tide.

lunar bulge Also termed lunar tide, it is the part of the tide caused by the gravitational attraction of the moon.

m

macrobenthos Organisms living on or in soft bottoms whose shortest dimension is greater than 0.5mm.

macroplankton Plant or animal plankton ranging in size from 2mm to 0.2mm.

magma Molten materials that form igneous rock upon cooling.

mangal The community of which mangroves are the dominant organisms.

mantle The earth's layer that is located between the core and the crust.

mechanical weathering The physical breakdown of rocks into finer particles.

media   The material immediately surrounding an organism—air or water.

megaplankton   Plant or animal plankton greater than 2mm in size.

meiobenthos   Organisms living on or in soft bottoms whose shortest dimension is between 0.5 and 0.1 mm.

meroplankton   Either phytoplankton or zooplankton that spend only a portion of their lives as plankton.

microbenthos   Organisms living on or in soft bottoms whose shortest dimension is less than 0.1mm.

microplankton   Plant or animal plankton ranging in size between 0.2 and 20μm.

midtide horizon   That area of the intertidal zone that is covered at high tide and exposed at low.

minor element   An element not required by all organisms but vital to the organisms in which it is found. Minor elements are present in the sea in concentrations that range from 1 part per thousand to 1 part per hundred.

mixed tide   A tide consisting of high and low waters that vary significantly in height per tidal day.

moraine   Large quantities of unsorted sediments deposited by a glacier.

mortality   Death rate. The death rate is calculated by dividing the number of deaths that occur during a given time interval by the average population.

n

nanoplankton   Phytoplankton or zooplankton that range between 20μm and 2μm in size.

natality   The number of births that occur within a given time period.

neap tide   The smaller than normal tidal variation that occurs when the moon and sun are in their out-of-phase position.

nekton   Organisms swimming within the water column.

neritic sediment   Mainly terrigenous sediments that have been transported to the sea by rivers and surface runoff.

neritic zone   A subdivision of the pelagic realm that encompasses all the water that lies above the continental shelf.

niche   The functional role of a population within its community.

node   The point on a standing wave at which vertical motion is minimal or absent.

nonpolar covalent bond   A chemical bond formed by the complete and equal sharing of electrons.

o

oceanic crust   A basalt crust that underlies the deep ocean offshore the continental slope.

oceanic zone   The subdivision of the pelagic realm that encompasses all the waters offshore the continental shelf.

ooze   Sediments that consist of over 30 percent of the remains of microscopic organisms.

outer core   The molten outer portion of the core of the earth.

outer neritic zone   The subdivision of the pelagic realm that encompasses all the waters of the continental shelf having a depth of 50 meters or more.

p

passive continental margin   Sites of geological inactivity.

pelagic realm   The water column.

pelagic sediment   Sediments found offshore the continental shelf on the deep-ocean floor.

perigee tide   The tide that is formed when the moon is at its closest to earth.

phytoplankton   Plant plankton.

picoplankton   Phytoplankton less than 0.2 microns in size.

plankton   Any organism, regardless of size, that is free-floating or drifting, with its movements controlled by the motion of the water.

plankton bloom   A high concentration of phytoplankton that occurs seasonally as a result of their high reproductive rate. Different species may bloom at different times.

plate tectonics   The theory and study of the formation, interaction, and destruction of the earth's lithospheric plates.

polar covalent bond   A chemical bond formed by the unequal sharing of electrons.

polar easterlies   Air masses that move from the poles in a northeasterly and southeasterly direction.

polychromatic light   Light containing all the colors of the electromagnetic spectrum.

population   A group of organisms of the same species inhabiting a given area.

pressure-gradient current   Movements induced by the different pressures on adjacent water masses.

primary bay-mouth bar   A bay-mouth bar growing in the same general direction as the longshore current.

primary coastline   Coasts that have recently come into contact with the sea.

proton   A positively charged particle found in the nucleus of an atom.

pseudocrystalline water   Water molecules between the temperatures of 4°C and 0°C that are in the process of forming the hexagonal ice crystal.

pteropod   A small free-swimming pelagic snail. Pteropod shells form oozes in some areas of the sea.

pteropod ooze   Sediments that are composed of over 30 percent pteropod shells.

r

radiolarian   A single-celled zooplankton having a skeleton of silica spicules.

radiolarian ooze   Sediments that consist of over 30 percent of the remains of radiolarians.

red tide   The red or reddish-brown discoloration of water caused by high concentrations of phytoplankton, primarily the dinoflagellates. Red tides occur most frequently in coastal areas.

reflection   A process that occurs when the moving energy form of a wave is forced back upon itself.

refraction   The slowing and concomitant bending of a portion of a wave in response to its moving over subsurface contours.

relict sediment   Sediment that is out of equilibrium with its present environment.

rift valley   A depression formed by faulting in an area of divergence.

rip current   A strong current that flows seaward perpendicular to the shoreline.

rise   A long, broad elevation that rises gently and smoothly from the sea floor.

s

salinity   The total amount of dissolved material that is present in 1 kg. of water, assuming that all the carbonates have been converted to oxides and the bromides and iodides replaced by chloride, with the organic substances oxidized.

salinity density barrier A barrier induced by the variations in the salinity of different vertical water masses.

salt-water wedge An intrusion of seawater in a tidal estuary or river. The seawater moves in the form of a wedge in response to density differences between the saline seawater and the less dense, less saline estuarine water.

sea-floor spreading The movement of plates away from the mid-ocean ridge or rise.

sea mount High abyssal hills.

seasonal succession The seasonal change in populations.

secondary bay-mouth bar A bay-mouth bar growing opposite the direction of the longshore current.

secondary coastline An area that has been shaped by marine processes.

sedimentary rock A rock formed by the consolidation of loose sediment.

seiche The rhythmic, back-and-forth motion of waves.

seismic low-velocity zone The asthenosphere.

seismic sea wave Waves generated by disturbances within the earth's crust (*see* tsunami).

seismic wave Waves of energy generated by earthquakes. These waves slow as they travel through the asthenosphere.

semidiurnal tide Two equal or nearly equal high and low tides per day.

sessile Immobile.

shallow water wave Any wave with a ratio of water depth to wave length of less than 1:25.

short-wave order Small aggregations of molecules held together by relatively weak intermolecular forces.

siliceous ooze Sediments that consist of over 30 percent silicon.

simple inlet An inlet with a channel perpendicular to the barrier beach.

sine wave A mathematical expression for a smooth, regular oscillation.

solar bulge Also termed solar tide, it is the part of the tide caused by the gravitational attraction of the sun.

spatial unavailability A condition in which necessary materials, such as nitrogen and phosphorous, are below the euphotic zone and are not available for use by plants.

species Any group of organisms that are actually or potentially capable of reproducing and producing viable, fertile offspring.

specific heat The amount of energy that is required to raise the temperature of 1 gram of a substance 1°C.

spit-built estuary An estuary protected by a spit.

spring tide A higher than normal tide caused when the sun and the moon are in their in-phase position.

stable shoreline A shoreline that is neither gaining nor losing sediment.

stenohaline organism An organism that has little tolerance for salinity variations.

subduction The sinking of a plate.

sublittoral zone A subdivision of the benthic realm. It encompasses the area immediately seaward of the low-tide horizon to the continental slope.

submergent shoreline A shoreline resulting from the sinking of coastal areas.

substrate The material on which or in which an organism lives.

subtidal zone The area immediately beyond the low-tide horizon.

surf zone The area between the shoreline and the breaking waves.

swash zone The area in which water moves up a beach after a wave breaks.

t

tectonic forces Forces generated within the earth.

temperature density barrier A barrier induced by temperature differentials between adjacent vertical water masses.

terrigenous sediment Sediments derived from the land.

test The protective outer covering of microscopic marine organisms.

thermocline An area where the temperature changes rapidly with depth.

thermohaline current Water movements in response to the density differences of water masses.

tidal bore The high-tide crest that advances up a funnel-shaped river or estuary.

tidal bulge A wave with a very long period, produced by the gravitational attraction of the moon and, when in-phase, the sun.

tidal current The alternating horizontal movement of water associated with the rise and fall of the tide.

tidal day The time encompassed by two successive passes of the moon over a given point. A tidal day is approximately 24 hours and 50 minutes in length.

tombolo The sediments that link islands to the mainland.

trace element An element not required by all organisms but vital to the organisms in which it is found. Trace elements are present in the sea in concentrations of less than 1 part per thousand.

trade wind The northeasterly and southeasterly movements of air masses in the Northern and Southern hemispheres, respectively. These air masses move from subtropical high-pressure belts toward the equator.

transform fault A fracture in the oceanic crust associated with the oceanic ridges.

transform-plate boundary A boundary formed by plates that move horizontally past each other.

trench A long, narrow, deep depression in the deep-ocean floor.

trophic level The level at which an organism obtains its food.

tsunami A seismic sea wave.

Type I coastline A coastline recently exposed to the sea.

Type II coastline A mid-aged coastline.

Type III coastline An old coastline.

typhoon *See* hurricane.

u

ultraplankton Phytoplankton or zooplankton less than $2\mu m$ in size.

upwelling The return of deep, cold water to the euphotic zone. Upwelling occurs when surface waters are moved offshore.

v

viscosity A liquid's resistance to flow.

w

water mass A body of water that may be identified by particular characteristics (generally temperature and/or salinity).

wave-cut terrace Horizontal platforms of consolidated rock formed in response to erosion of a cliff face.

wave frequency The number of waves passing a fixed point in a given time.

wave group A series of waves in which the wave direction, wave length, and wave height are similar.

wave height The vertical distance from a wave crest to its trough.

wave length The horizontal distance between adjacent wave crests.

wave period The amount of time required for one wave crest to travel one wave length.

wave speed The relationship of wave length to wave period.

wave steepness The ratio of wave length to wave height.

wave train A series of waves moving from the same direction at the same speed.

weathering The process by which rocks are broken down by thermal, chemical, or mechanical processes.

westerlies Air masses moving from the west between latitudes 30° to 60° north and 30° to 60° south. They move in a southeasterly direction in the Northern Hemisphere and a northwesterly direction in the Southern Hemisphere.

wind-drift current A current induced by the wind.

wind wave Waves generated by the wind.

Z

zone Any area that encompasses a defined feature, structure, or property.

zooplankton Animal plankton.

Photos

Chapter 1
p. 3(top): NASA; **p. 3(bottom):** © T. Leahy; **p. 5:** U.S. Geological Survey; **p. 7(top):** © Tom McHugh/Photo Researchers, Inc.; **p. 7(bottom):** © Russ Kinne/ Photo Researchers, Inc.

Chapter 2
p. 13: NASA; **p. 16:** AP/Wide World Photos; **p. 17:** © R. J. Rogerson, Memorial University of Newfoundland; **p. 19:** © B. F. Molnia, TERRAPHOTOGRAPHICS/BPS; **p. 20:** Scripps Institution of Oceanography; **p. 21(both):** EROS Data Center; **p. 22(top):** NASA; **p. 22(bottom):** EROS Data Center; **p. 25:** © Rod Catanach, WHOI 1981; **p. 27:** NASA; **p. 30(top):** © B. F. Molnia, TERRAPHOTOGRAPHICS/BPS; **p. 30(bottom):** © John A. Black

Chapter 3
p. 40(top): © B. F. Molnia, TERRAPHOTOGRAPHICS/BPS; **p. 40(bottom), p. 41:** © John A. Black; **p. 42:** © Ray Welch; **p. 43(left):** © William Ormerod/ New England Aquarium; **p. 43(right):** James L. Sumich; **p. 44:** © John A. Black; **p. 45:** © D. J. Cross/BPS; **p. 47:** Lamont-Doherty Geological Observatory; **p. 49:** EROS Data Center

Chapter 4
p. 53: United States Coast Guard; **p. 58(top):** © B. F. Molnia, TERRAPHOTOGRAPHICS/BPS; **p. 58(bottom), p. 59:** © John A. Black; **p. 60:** Lockwood, Keeslar and Bartlett, Inc.; **p. 61:** © John D. Cunningham/Visuals Unlimited; **p. 63:** Lockwood, Keeslar and Bartlett, Inc.; **p. 67:** © John A. Black; **p. 69:** © Gail L. Marquardt; **p. 70:** N.O.A.A.

Chapter 5
p. 76: © John A. Black; **p. 77:** © B. F. Molnia, TERRAPHOTOGRAPHICS/BPS; **p. 80:** Lockwood, Keeslar and Bartlett, Inc.; **p. 81:** © Gail L. Marquardt; **p. 82:** EROS Data Center; **p. 83:** Lockwood, Keeslar and Bartlett, Inc.; **p. 84:** © John A. Black; **p. 85(left):** James L. Sumich; **p. 85(right):** © S. K. Webster, Monterey Bay Aquarium/BPS; **p. 86:** U.S. Geological Survey; **p. 88:** NASA; **pp. 89, 90:** EROS Data Center; **p. 91:** N.O.A.A./U.S.G.S. EROS Data Center

Chapter 7
p. 111: EROS Data Center; **p. 114:** © Fred Ward/Black Star; **p. 117:** From Plummer, Charles C., and McGeary, David, PHYSICAL GEOLOGY, 3d ed., Wm. C. Brown Publishers; **p. 120:** © John A. Black; **p. 122:** © Karl H. Maslowski/Photo Researchers, Inc.; **p. 123:** Great Gull Project, American Museum of Natural History.

Chapter 8
p. 129: © Claudia Mills, University of Washington/BPS; **p. 131(top):** © William J. Weber/Visuals Unlimited; **p. 131(bottom):** © William Ormerod/New England Aquarium; **p. 132:** © John D. Cunningham/Visuals Unlimited; **pp. 133, 135, 136:** © R. J. Stone; **p. 144:** N.O.A.A.; **p. 146(left):** Scripps Institution of Oceanography, University of California, San Diego; **p. 146(right):** © Stan W. Elems/ Visuals Unlimited

Chapter 9
p. 149: © B. F. Molnia, TERRAPHOTOGRAPHICS/BPS; **p. 151:** © B. J. Miller, Fairfax, VA/ BPS; **p. 154:** © John D. Cunningham/Visuals Unlimited; **p. 155(both):** N.O.A.A.; **p. 164:** © Richard Ellis/Photo Researchers, Inc.; **p. 165:** © H. W. Pratt/BPS; **p. 166:** © Stan W. Elems/Visuals Unlimited; **p. 167:** N.O.A.A.

Chapter 10
p. 187(top): © J.N.A. Lott, McMaster University/BPS; **p. 187(bottom):** © William Ormerod/Nature Photography

Chapter 11
p. 195(both): © John A. Black; **p. 199:** N.O.A.A./U.S.G.S. EROS Data Center; **p. 201:** © C. R. Wyttenbach, University of Kansas/BPS; **p. 202:** © Jack Dermid/Photo Researchers, Inc.

Chapter 12
p. 205: © John D. Cunningham/ Visuals Unlimited; **p. 206:** © Gail L. Marquardt; **p. 207:** © John A. Black; **pp. 209, 210, 212, 213:** Lockwood, Keeslar and Bartlett, Inc.; **p. 214(top):** NASA; **p. 214(bottom):** EROS Data Center; **p. 215, 217(both):** Lockwood, Keeslar and Bartlett, Inc.; **p. 219:** © John A. Black; **p. 220:** N.O.A.A./U.S.G.S. EROS Data Center; **p. 221:** © Gail L. Marquardt; **p. 222(top):** © John A. Black; **p. 222(bottom):** © B. F. Molnia, TERRAPHOTOGRAPHICS/ BPS; **p. 223:** Lockwood, Keeslar and Bartlett, Inc.; **pp. 224, 225:** © B. F. Molnia, TERRAPHOTOGRAPHICS/BPS; **pp. 227, 228:** Lockwood, Keeslar and Bartlett, Inc.; **p. 229:** © T. Leahy

Credits

Chapter 13

p. 233: © H. W. Pratt/BPS;
p. 234: © Ray Welch; **p. 235:**
© John A. Black; **p. 237:** © Elgin
Ciampi/Photo Researchers, Inc.;
p. 244(left): James L. Sumich;
p. 244(right): © John A. Black;
p. 245(top): © Gail L. Marquardt;
p. 245(bottom): © Bart Cadbury/
Photo Researchers, Inc.;
p. 246(top): © John A. Black;
p. 246(bottom): © Gail L.
Marquardt; **p. 247:** © John A.
Black; **p. 248:** © J. Robert
Waaland/BPS; **p. 249:** © H. W.
Pratt/BPS; **p. 250:** © Gail L.
Marquardt; **p. 252:** © Stan E.
Elems/Visuals Unlimited; **pp. 254,
255(top left):** © B. J. Miller,
Fairfax, VA/BPS; **p. 255(bottom
left):** © Gail L. Marquardt;
p. 255(right): © S. K. Webster,
Monterey Bay Aquarium/BPS;
p. 256: © Gail L. Marquardt;
p. 257: © John A. Black

Chapter 14

p. 263(both): © H. W. Pratt/BPS;
p. 264: © Tom McHugh/Photo
Researchers, Inc.; **p. 266:**
© Alexander Lowry/Photo
Researchers, Inc.; **p. 267(top):**
© R. Humbert/BPS;
p. 267(bottom): © S. K. Webster,
Monterey Bay Aquarium/BPS;
p. 268: © Tom McHugh/Photo
Researchers, Inc.; **p. 269:**
© William Ormerod/New England
Aquarium; **p. 272:** © John A.
Black

Illustrations

Chapter 1

Figs. 1.1–1.3, 1.5, TA 1.4, TA 1.7:
Norm Frisch. **Fig. 1.4:** From R. S.
Dietz and J. C. Holden, *Journal
of Geophysical Research,* V. 75,
1970. American Geophysical
Union, Washington, D.C. **Fig. 1.6:**
Fine Line Illustrations, Inc.

Chapter 2

**Figs. 2.1–2.7, 2.9, 2.11,
2.13–2.19, 2.21:** Norm Frisch.
Figs. 2.12, 2.23, 2.25: From
Plummer, Charles C., and David
McGeary, *Physical Geology 3d
ed.* © 1979, 1982, 1985 Wm. C.
Brown Publishers, Dubuque,
Iowa. All Rights Reserved.
Reprinted by permission. **Figs.
2.20, 2.22, 2.24:** From R. S. Dietz
and J. C. Holden, *Journal of
Geophysical Research,* V. 75,
1970. American Geophysical
Union, Washington, D.C.

Chapter 3

Figs. 3.1, 3.2, 3.4: Norm Frisch.

Chapter 4

Figs. 4.1–4.23: Norm Frisch.

Chapter 5

Figs. 5.1–5.11: Norm Frisch.

Chapter 6

Figs. 6.1–6.6, 6.8–6.24: Fine Line
Illustrations, Inc. **Fig. 6.7:** From
Mader, Sylvia S., *Inquiry into Life
4th ed.* © 1976, 1979, 1982, 1985
Wm. C. Brown Publishers,
Dubuque, Iowa. All Rights
Reserved. Reprinted by
permission.

Chapter 7

Figs. 7.1, 7.6–7.8, TA 7.5: Norm
Frisch. **Figs. 7.2–7.5, 7.9:** Fine
Line Illustrations, Inc.

Chapter 8

Figs. 8.1, 8.3–8.5: Fine Line
Illustrations, Inc. **Fig. 8.2:** From
Sumich, James L., *An
Introduction to the Biology of
Marine Life 3d ed.* © 1976, 1980,
1984 Wm. C. Brown Publishers,
Dubuque, Iowa. All Rights
Reserved. Reprinted by
permission. **Figs. 8.6–8.8:** Norm
Frisch.

Chapter 9

Figs. 9.1, 9.3–9.5, 9.8, 9.9: Fine
Line Illustrations, Inc. **Figs. 9.2,
9.6, 9.7, 9.10–9.14:** Norm Frisch.

Chapter 10

Figs. 10.1–10.26: Norm Frisch.

Chapter 11

**Figs. 11.1–11.11, TA 11.2, TA
11.3:** Norm Frisch.

Chapter 12

Figs. 12.1–12.14: Norm Frisch.

Chapter 13

Figs. 13.1, 13.4–13.9: Norm
Frisch. **Figs. 13.2, 13.3:** Fine Line
Illustrations, Inc.

Chapter 14

Figs. 14.1–14.3: Norm Frisch.
Figs. 14.4–14.6: Fine Line
Illustrations, Inc.

Index